Lecture Notes in Economics and Mathematical Systems

Managing Editors: M. Beckmann and H. P. Künzi

186

Ching-Lai Hwang
Kwangsun Yoon

Multiple Attribute Decision Making
Methods and Applications
A State-of-the-Art Survey

Springer-Verlag
Berlin Heidelberg New York 1981

Managing Editors

Prof. Dr. M. Beckmann
Brown University
Providence, RI 02912/USA

Prof. Dr. H. P. Künzi
Universität Zürich
CH-8090 Zürich/Schweiz

Authors

Ching-Lai Hwang
Dept. of Industrial Engineering
Durland Hall
Kansas State University
Manhattan, KS 66506/USA

Kwangsun Yoon
Dept. of Industrial Engineering
and Management Science
Fairleigh Dickinson University
Teaneck, NJ 07666/USA

AMS Subject Classifications (1970): 90-02, 90 B 99, 90 C 99

ISBN 3-540-10558-1 Springer-Verlag Berlin Heidelberg New York
ISBN 0-387-10558-1 Springer-Verlag New York Heidelberg Berlin

Printing and binding: Beltz Offsetdruck, Hemsbach/Bergstr.
2142/3140-543210

PREFACE

This monograph is intended for an advanced undergraduate or graduate course as well as for the researchers who want a compilation of developments in this rapidly growing field of operations research. This is a sequel to our previous work entitled "Multiple Objective Decision Making--Methods and Applications: A State-of-the-Art Survey," (No. 164 of the Lecture Notes).

The literature on methods and applications of Multiple Attribute Decision Making (MADM) has been reviewed and classified systematically. This study provides readers with a capsule look into the existing methods, their characteristics, and applicability to analysis of MADM problems.

The basic MADM concepts are defined and a standard notation is introduced in Part II. Also introduced are foundations such as models for MADM, transformation of attributes, fuzzy decision rules, and methods for assessing weight.

A system of classifying seventeen major MADM methods is presented. These methods have been proposed by researchers in diversified disciplines; half of them are classical ones, but the other half have appeared recently. The basic concept, the computational procedure, and the characteristics of each of these methods are presented concisely in Part III. The computational procedure of each method is illustrated by solving a simple numerical example.

Part IV of the survey deals with the applications of these MADM methods. The literature has been classified into selection of commodity, site, people, project, public facility, etc. A summary of each reference on applications is given.

A choice rule for MADM methods, a unified approach to MADM problems, and proposed future study are presented in Part V.

An updated bibliographical listing of twenty-five books, monographs or conference proceedings, and about 500 selected papers, reports or theses is presented.

We are indebted to the outstanding pioneering survey of this field done by Dr. Kenneth R. MacCrimmon in 1968 and 1973; and to Professors Doris Grosh, William Schenck-Hamlin and P. L. Yu for their various comments and suggestions. The first draft was used in the first author's Spring 1980 class of "Advanced Topics in Operations Research." Dale G. Finkner, M. H. Lee, M. J. Lin, Cynthia S. McCahon, K. S. Raju, and Larry M. Strecker have tested and critically evaluated many methods. Special thanks are due to Merla Oppy for typing and Jean Burnham for editing.

This study was partly supported by the Office of Naval Research, and Department of Energy.

C. L. Hwang
Kansas State University
Manhattan, Kansas
Fall 1980

Kwangsun Yoon
Fairleigh Dickinson University
Teaneck, New Jersey
Fall 1980

TABLE OF CONTENTS

LIST OF FIGURES

LIST OF TABLES

I. INTRODUCTION

Multiple criteria decision making (MCDM) refers to making decisions
in the presence of multiple, usually conflicting, criteria. Problems for multiple
criteria decision making are common occurrences in everyday life. For example:

In a personal context, the job one chooses may depend upon its prestige,
location, salary, advancement opportunities, working conditions, and so on.
The car one buys may be characterized in terms of price, gas mileage, style,
safety, comfort, etc. A young man/woman may choose a wife/husband based on
her/his intelligence, looks, character, etc.

In a business context, a business executive's choice of corporate strategy
may depend on the company's earnings over a period of time, its stock price,
share of market, goodwill, labor relations, corporate image, obligation to
society, and so forth. Automobile manufacturers in Detroit want to design a
model which maximizes fuel efficiency, maximizes riding comfort, minimizes
production cost, etc.

In an academic context, a university administrator's selection of the
future configurations of the university would be based on number of regular
faculty, number of auxiliary faculty, undergraduate enrollment, graduate enrollment, tuition level, faculty leverage, new programs, and net improvement, etc.

In a public context, the water resources development plan for a community
should be evaluated in terms of cost, probability of water shortage, energy
(reuse factor), recreation, flood protection, land and forest use, water
quality, etc.

In a government context, the Department of Transportation should devise a transportation system which would minimize travel time, departure delays, arrival delays, fare cost, and so on. The choice of missile systems for the Air Force would be based on speed, yield, accuracy, range, vulnerability, reliability, etc.

The problems of MCDM are widely diverse. However, even with the diversity, all the problems which are considered here share the following common characteristics:

Multiple objectives/attributes. Each problem has multiple objectives/attributes. A decision maker must generate relevant objectives/attributes for each problem setting.

Conflict among criteria. Multiple criteria usually conflict with each other. For example, in designing a car, the objective of higher gas mileage might reduce the comfort rating due to the smaller passenger space.

Incommensurable units. Each objective/attribute has a different unit of measurement. In the car selection case, gas mileage is expressed by miles per gallon (MPG), comfort is by cu ft if it is measured by passenger space, safety may be indicated in a nonnumerical way, cost is indicated by dollars, etc.

Design/selection. Solutions to these problems are either to design the best alternative or to select the best one among previously specified finite alternatives. The MCDM process involves designing/searching for an alternative that is the most attractive over all criteria (dimensions).

One may notice that there exist two alternative sets due to the different problem settings: one set contains a finite number of elements (alternatives), and the other has an infinite number. For instance a car a customer may purchase (select) is among the available finite models auto companies have produced; but a model which a certain company mass produced is among the infinite number of options which engineers may have designed.

The problems of MCDM can be broadly classified into two categories in this respect: Multiple Attribute Decision Making (MADM), and Multiple Objective Decision Making (MODM). In actual practice this classification is well fitted to the two facets of problem solving--MADM is for selection (evaluation), MODM is for design. This is a widely accepted classification [BM-5, BM-11, 275, 283]. Table 1.1 shows the contrast of the features between these two classes.

Multiple objective decision making (MODM) is not associated with the problem where the alternatives are predetermined. The thrust of these models is to design the 'best' alternative by considering the various interactions within the design constraints which best satisfy the DM by way of attaining some acceptable levels of a set of some quantifiable objectives. The common characteristics of MODM methods are that they possess: (1) a set of quantifiable objectives; (2) a set of well defined constraints, (3) a process of obtaining some tradeoff information, implicit or explicit, between the stated quantifiable objectives and also between stated or unstated nonquantifiable objectives. Thus MODM is associated with design problems (in contrast to selection problems for the MADM).

Literature on MODM methods and applications has been reviewed extensively by Hwang and Masud [BM-11].

The distinguishing feature of the MADM is that there is usually a limited (and countably small) number of predetermined alternatives. The alternatives have associated with them a level of the achievement of the attributes (which may not necessarily be quantifiable) based on which the final decision is to be made. The final selection of the alternative is made with the help of inter- and intra-attribute comparisons. The comparisons may involve explicit or implicit tradeoff.

Table 1.1   MADM vs MODM

	MADM	MODM
Criteria (defined by)	Attributes	Objectives
Objective	Implicit (ill defined)	Explicit
Attribute	Explicit	Implicit
Constraint	Inactive (incorporated into attributes)	Active
Alternative	Finite number, discrete (prescribed)	Infinite number, continuous (emerging as process goes)
Interaction with DM	Not much	Mostly
Usage	Selection/Evaluation	Design

Previous Research on MADM

Although the effort to introduce the concept of multiple criteria into the normative decision making process started within the last two decades, the study on multiple criteria has a long tradition. We notice the earlier works by the researchers in many disciplines: management science, economics, psychometrics, marketing research, applied statistics, decision theory, and so on. They confront multiple criteria in quite different situations. Consequently each area has developed method(s) for its own particular usage. For example:

Decision theory. Maximin, prior probability, utility theory.

Economics. Pareto optimality, von Neumann-Morgenstern utility, social welfare function, benefit/cost analysis.

Statistics. Multivariate regression, discrepancy analysis, factor analysis.

Psychometrics. Multidimensional scaling, conjoint measurement.

Their orientation and motivation is mostly to explain, rationalize, understand, or predict decision behavior--not to guide the decision making. If we define MADM in a narrow sense as decision aids to help a DM identify the best alternative that maximizes his/her satisfaction with respect to more than one attribute, many of the above approaches may not be directly used in MADM situations. But it should be recognized that the above methods/theories have been the basis in the development of MADM. The substantial advancement of MADM has been made in the last two decades.

It was Churchman, Ackoff, and Arnoff [57] in 1957 who first treated a MADM problem (selecting business investment policy) formally using simple additive weighting method. Many potentially useful concepts/methods for MADM had been laid aside until 1968 when MacCrimmon [273] reviewed the methods and the applications of the MADM. He classified his collection of ten methods according to the number of dimensionality (single, intermediate, full) of attribute treatment.

In his second review [275] in 1973, more methods are added and grouped according to the structure of method, compensatoriness, preference input, etc. It is rather surprising that little effort has been given on the review of MADM since then. The literature on MADM is briefly treated in some of the MCDM reviews such as   [304, 385].

We have seen rapid theoretical development in multiattribute utility theory (MAUT) which is a solution approach of MADM under uncertainty. It started with simple additive utility and has gone to quasi-additive (multiplicative form) and multilinear utility function. There are several reviews on this field: Fishburn [120, 126], Huber [196, 197], and Farquhar [98]'s reviews on MAUT are exhaustive and theoretically in-depth. Keeney and Raiffa [BM-13] published a voluminous text basically on MAUT.

Multiple criteria decision making problems have been presented in books and monographs. Table 1.2 shows the classification. Several symposium proceedings and collected papers presenting recent progress (1973-1978) in theory, method, and applications on MCDM have been edited by Cochrane and Zeleny [BM-4], Leitman and Marzollo [BM-16], Wendt and Vleck [BM-22], Thiriez and Zionts [BM-21], Zeleny [BM-24], Bell, Keeney and Raiffa [BM-3], Starr and Zeleny [BM-20], and Zionts [BM-24]. However, only a small number of articles in these volumes deal with MADM problems except on MAUT.

Barret [BM-1] and Easton [BM-6] published texts on MADM. Especially on multidimensional scaling there are several books available: Bechtel [BM-2], Green et al. [BM-9, BM-10], Krushkal [BM-14], Shepard [BM-19]. Nijkamp and van Delft [BM-18] have a monograph on ELECTRE. A recent text by Nijkamp [BM-17a] covers both MODM and MADM areas.

Objectives of the Present MADM Review

The rapid progress of MADM in recent years makes necessary a thorough review of the existing literature and a systematic classification of methods. In addition to reviewing the methods, we will also review the actual or proposed applications of these methods. Some basic concepts and terminology will be defined so that we

Table 1.2 Classification of Books, Monographs and Conference Proceedings

Class	References
A. Directly related to the topic	
Books	BM-1, BM-2, BM-6, BM-8, BM-9, BM-10, BM-13
Monographs	BM-7, BM-14, BM-17, BM-18, BM-17a
Edited conference proceedings	BM-3, BM-4, BM-15, BM-16, BM-19, BM-20, BM-21, BM-22, BM-23, BM-24
B. Indirectly related to the topic	BM-5, BM-11, BM-12

can explain the literature with a unified notation of the most used terms.

This review will provide readers with a capsule look into the existing methods, their characteristics, and applicability to analyzing MADM problems.

Classification of MADM Methods

The decision makers' judgments vary in form and depth. One may not indicate his preferences at all, or may represent his preference through the form of attribute or alternative. The degree of judgment skill also varies. For instance, we may list the different preference information on attributes by the ascending order of complexity: standard level, ordinal, cardinal, and marginal rate of substitution. MADM methods are introduced to meet these various situational judgments. We classify methods for MADM based upon different forms of preference information from a DM. A taxonomy of MADM methods is shown in Fig. 1.1. The classification has been made in three stages: Stage I: the type of information (attribute or alternative or neither) needed from the DM; Stage II: the salient feature of the information needed; Stage III: the major methods in any branch formed from Stages I and II. Finally in Table 1.3, all references are classified according to the methods in Step III.

Before going into the actual review, some key concepts and notations will be defined in Part II so that we can explain the literature with a unified notation of the most used terms. Also in Part II some supporting techniques for MADM, such as transformation of attributes and assessment of attribute weights, are discussed. In Part III, the literature dealing with the techniques of MADM will be reviewed and classified. The computational procedure of each method is illustrated by solving a few simple numerical examples. In Part IV, the literature dealing with the actual or proposed applications of MADM methods will be reviewed. Finally a bibliography of works related to MADM is given.

9

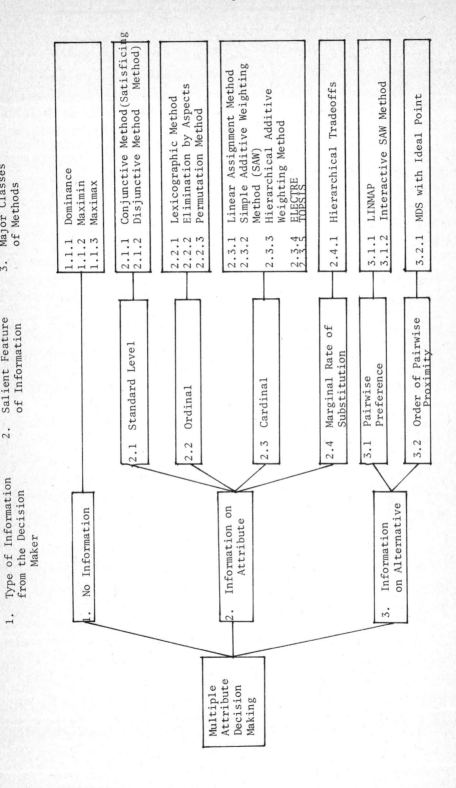

Fig. 1.1 A taxonomy of methods for multiple attribute decision making.

Table 1.3   Classification of references on multiple attribute decision making.

CLASS	REFERENCES
1. Methods of No Information Given:	
1.1.1 Dominance	23, 44, 163, 400, 426, 443, 444, 445
1.1.2 Maximin	19, 133, 273, 436, 437, 440
1.1.3 Maximax	70, 133, 273, 299, 178a
2. Methods for Information on Attribute Given:	
2.1. Methods for Standard Level of Attribute Given	
2.1.1 Disjunctive Method (Satisficing Method) 2.1.2 Conjunctive Method	3, 14, 70, 88, 240, 311, 319, 334, 373, 406
2.2. Methods for Ordinal Information of Attributes Given	
2.2.1 Lexicographic Method	26, 75, 93, 120, 191, 269, 322, 409
2.2.2 Elimination by Aspect	28, 410, 410a, 410b
2.2.3 Permutation Method	315
2.3. Methods for Cardinal Information of Attribute Given	
2.3.1 Linear Assignment Method	25, 35
2.3.2 Simple Additive Weighting Method	1, 56, 62, 85, 86, 87, 90a, 91, 108, 127, 151, 152, 153, 154, 155, 156, 214, 239a, 267, 268, 309, 356, 358, 416
2.3.3 Hierarchical Additive Weighting Method	51, 350, 374, 401
2.3.4 ELECTRE	67, 158, 169, 205, 206, 264, 297, 303, 345, 347, 348, 349, 413
2.3.5 TOPSIS	66, 440a, 440b

Table 1. 3 (continued)

Because of the broad interdisciplinary character of this field, the literature is diversified in many journals as shown in Table 1.4. Journals in which MADM articles appear frequently are indicated in the list.

Although we have tried to give a reasonably complete survey, those papers not included were either inadvertently overlooked or considered not to bear directly on the topics in this survey. We apologize to both the readers and the researchers if we have omitted any relevant papers.

Table 1.4   List of Journals Relevant to MADM

1. Academy of Management Jol.
2. American Economic Review
3. American Jol. of Psychology
4. American Jol. of Sociology
5. American Political Science Review

6. American Psychologist
7. American Sociological Review
8* Annals of Economic & Social Measurement
9. Annals of Mathematical Statistics
10. Annals of Statistics

11. Annual Review of Psychology
12. Applied Economics
13* Behavioral Science
14* Bell Jol. of Economics & Management Science
15. Biometrica

16. Bulletin of the American Mathematical Society
17. Cognitive Psychology
18* Computers & Operations Research, An International Journal
19* Decision Sciences
20* Econometrica

21. Economica
22. Economic Jol.
23* Engineering Economist
24* Harvard Business Review
25. Human Factors

26. IEEE Transactions on Engineering Management
27. IEEE Transactions on Reliability
28* IEEE Transactions on Systems Science and Cybernetics
29. International Economic Review
30. International Encyclopedia of the Social Science

31. Israel Jol. of Mathematics
32. Jol. of Abnormal & Social Psychology
33. Jol. of Accounting Research
34. Jol. of Advertising Research
35* Jol. of American Institute of Planners

36* Jol. of American Statistical Association
37. Jol. of American Statistical Society
38. Jol. of Business
39* Jol. of Consumer Research
40. Jol. of Economic Theory

---

*Indicates MADM articles appear frequently in this journal.

41. Jol. of Experimental Psychology
42. Jol. of Finance
43. Jol. of Financial and Qualitative Analysis
44. Jol. of Financial Economics
45. Jol. of Industrial Engineering

46. Jol. of Marketing
47* Jol. of Marketing Research
48. Jol. of Mathematical Analysis and Applications
49. Jol. of Mathematical Economics
50. Jol. of Mathematical Psychology

51* Jol. of Optimization Theory & Applications
52. Jol. of Personality and Social Psychology
53. Jol. of Political Economy
54. Jol. of Purchasing
55* Jol. of Regional Science

56. Jol. of Social and Clinical Psychology
57. Jol. of Social Psychology
58. Jol. of Symbolic Logic
59* Jol. of the Market Research Society
60. Jol. of the Royal Statistical Society

61* Management Science
62. Mathematical Programming
63. Mathematics of Operations Research
64. Multivariate Behavioral Research
65. Naval Research Logistics Quarterly

66* OMEGA
67* Operations Research
68* Operations Research Quarterly
69* Organizational Behavior & Human Performance
70* Papers, Regional Science Association

71. Philosophy of Science
72. Psychological Bulletin
73. Psychological Review
74* Psychometrika
75. Psychonomic Science

76. Public Administration Review
77. Quarterly Jol. of Economics
78. Review of Economics and Statistics
79. Review of Economic Studies
80. SIAM Jol. on Applied Mathematics

81* Sloan Management Review
82. Social Indicators Research
83. Southern Economic Jol.
84. Synthese
85. The Quarterly Jol. of Economics

---

* Indicates MADM articles appear frequently in this journal.

86. The Review of Economic Studies
87* TIME Studies in the Management Science
88. Transfusion
89. Transactions of the American Mathematical Society
90. Urban Analysis

91* Water Resources Bulletin
92* Water Resources Research

---

*Indicates MADM articles appear frequently in this journal.

II. BASIC CONCEPTS AND FOUNDATIONS

1. DEFINITIONS

The four words most used in MCDM literature are: attributes, objectives, goals and criteria. There are no universal definitions of these terms [BM-13]. Some authors make distinctions in their usage while many use them interchangeably. We will make some distinctions among these words in terms of their usage.

1.1. Terms for MCDM Environment

Criteria: A criterion is a measure of effectiveness. It is the basis for evaluation. Criteria are emerging as a form of attributes or objectives in the actual problem setting.

Goals: Goals (synonymous with targets) are a priori values or levels of aspiration. These are to be either achieved or surpassed or not exceeded. Often we refer to them as constraints because they are designed to limit and restrict the alternative set. For example, the standard gas mileage, say 20 miles/gallon, set up by the federal government for 1980 models, is a constraint, whereas 30 miles/gallon may serve as a goal for the car manufacturer.

Attributes: Performance parameters, components, factors, characteristics, and properties are synonyms for attributes. An attribute should provide a means of evaluating the levels of an objective. Each alternative can be characterized by a number of attributes (chosen by DM's conception of criteria), i.e., gas mileage, purchasing cost, horsepower, etc. of a car.

Objectives: An objective is something to be pursued to its fullest. For example, a car manufacturer may want to maximize gas mileage or minimize production cost or minimize its level of air pollution. An objective generally indicates the direction of change desired.

Decision matrix: A MADM problem can be concisely expressed in a matrix format. A decision matrix D is a (mxn) matrix whose element x_{ij}'s indicate evaluation or value of alternative i, A_i, with respect to attribute j, X_j. Hence A_i, i = 1,2,...,m is denoted by

$$\underline{x}_i = (x_{i1}, x_{i2}, \ldots, x_{in})$$

and the column vector

$$\underline{x}_j = (x_{1j}, x_{2j}, \ldots, x_{mj})^T$$

shows the contrast of each alternative with respect to attribute j, X_j.

Decision matrix [44, 327] is also called goal achievement matrix [180, 182, 416], or project impact matrix [305].

Numerical Example (A Fighter Aircraft Selection Problem):

A country decided to purchase a fleet of jet fighters from the U.S. The Pentagon officials offered the characteristic information of four models which may be sold to that country. The Air Force analyst team of that country agreed that six characteristics (attributes) should be considered. They are: maximum speed (X_1), ferry range (X_2), maximum payload (X_3), purchasing cost (X_4), reliability (X_5), and maneuverability (X_6). The values of the six attributes for each model (alternative) are given in Table 2.1.

The decision matrix for the fighter aircraft selection problem, then, is:

	X_1	X_2	X_3	X_4	X_5	X_6
A_1	2.0	1500	20000	5.5	average	very high
A_2	2.5	2700	18000	6.5	low	average
A_3	1.8	2000	21000	4.5	high	high
A_4	2.2	1800	20000	5.0	average	average

D =

Table 2.1   A fighter aircraft selection problem

Alternatives	Attributes (X_j)					
	Maximum speed	Ferry range	Maximum payload	Acquisition cost	Reliability	Maneuverability
(A_i)	(Mach)	(NM)	(pounds)	($\$ \times 10^6$)	(high-low)	(high-low)
A_1	2.0	1500	20000	5.5	average	very high
A_2	2.5	2700	18000	6.5	low	average
A_3	1.8	2000	21000	4.5	high	high
A_4	2.2	1800	20000	5.0	average	average

1.2. MCDM Solutions

An optimal solution: An optimal solution to a MCDM problem is one which results in the maximum value of each of the objective functions simultaneously. That is, $\underline{x}^*$ is an optimal solution to the problem if $\underline{x}^* \ \epsilon \ X$ and $\underline{f}(\underline{x}^*) \geq \underline{f}(\underline{x})$ for all $\underline{x} \ \epsilon \ X$.

Since it is the nature of MCDM problems to have conflicting objectives/ attributes, usually there is no optimal solution to a MCDM problem. Looking at examples:

Example 1:

    Max $f_1 = x_1 + x_2$

    Max $f_2 = x_2 - x_1$

    subject to $x_1 \leq 3$

                      $x_2 \leq 3$

                      $x_1, x_2 \geq 0$

Example 2:

    Max $f_1 = x_2$

    Max $f_2 = x_2 - x_1$

    subject to $x_1 \leq 3$

                      $x_2 \leq 3$

                      $x_1, x_2 \geq 0$

Figure 2.1 shows the decision variable space representation of both examples. Figure 2.2 shows the objective function space representation of Example 1; let us call this S_1. Figure 2.3 shows the objective function space representation of Example 2; let us call it S_2.

We notice that Example 2 has an optimal solution. This optimal solution is located at $(x_1, x_2) = (0, 3)$ where $(f_1, f_2) = (3, 3)$; this is shown by point B in Fig. 2.3. For Example 1, there is no optimal solution. The maximum value of $f_1 = 6$ which occurs at $(x_1, x_2) = (3, 3)$; the maximum values of $f_2 = 3$ which occurs at $(x_1, x_2) = (0, 3)$. Point F in Fig. 2.2 gives $f_1 = 6$ and $f_2 = 3$, however, the point is outside the feasible region S_1.

It may be noted that the optimal solution is also known as the superior solution or the maximum solution.

An ideal solution: Optimal solution, superior solution, or utopia point [442] are equivalent terms indicating an ideal solution. In a MODM problem: $\max\limits_{x \in X} [f_1(x), f_2(x), \ldots, f_k(x)]$, $X = \{ x | g_i(x) \leq 0, i = 1,2,\ldots,m\}$, the ideal solution is the one that optimizes each objective function simultaneously, i.e., $\max\limits_{x \in X} f_j(x)$, $j = 1,2,\ldots,k$. An ideal solution can be defined as

$$A^* = (f_1^*, f_2^*, \ldots, f_j^*, \ldots, f_k^*)$$

where f_j^* is a feasible and optimal value for the jth objective function. This solution is generally infeasible; if it were not, then there would be no conflict among objectives.

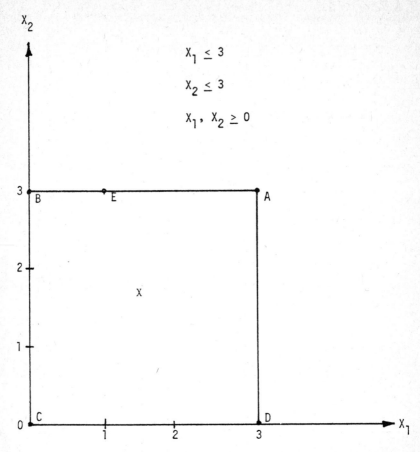

Fig. 2.1 Decision variable space representation of the feasible area of Examples 1 and 2.

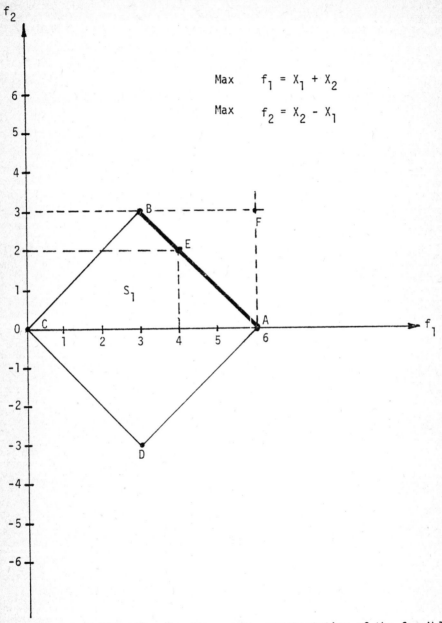

Fig. 2.2 Objective function space representation of the feasible area of Example 1.

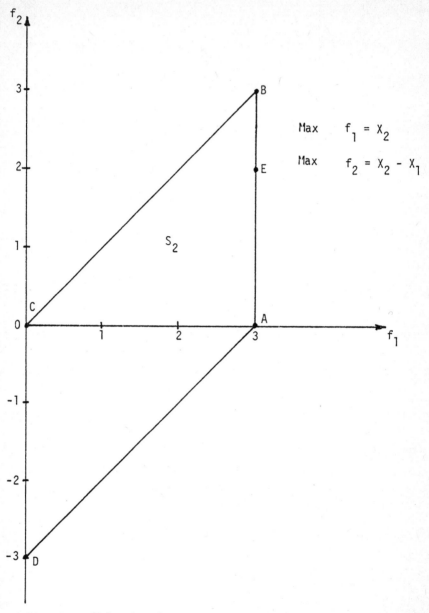

Fig. 2.3 Objective function space representation of the feasible
area of Example 2.

In MADM it is a hypothetical alternative whose Cartesian product is composed of the most preferable values from each attributes given in the decision matrix. Formally,

$$A^* = (x_1^*, x_2^*, \ldots, x_j^*, \ldots, x_n^*)$$

where

$$x_j^* = \max_i U_j(x_{ij}), \qquad i = 1,2,\ldots,m$$

where $U_j(\cdot)$ indicates the value/utility function of the jth attribute. It should be noted that an ideal solution in MADM is a subjective one whereas it is an objective one in a MODM environment. Hence locating an ideal solution is one of the topics in MADM study if a DM uses nonmonotonic value/utility functions.

Though an ideal solution does not actually exist, the concept of an ideal solution is essential in the development of MCDM methods. For example, a compromise model is based on the idea of obtaining a solution (alternative) which is close to the ideal solution.

What is then the best solution a DM is supposed to get? It is found in the set of nondominated solutions.

Nondominated solutions: This solution is named differently by different disciplines: nondominated solution, noninferior solution, and efficient solution in MCDM, a set of admissible alternatives in statistical decision theory, and Pareto-optimal solution in economics.

A feasible solution in MCDM is nondominated if there exists no other feasible solution that will yield an improvement in one objective/attribute without causing a degradation in at least one other objective/attribute.

For Example 1, all solutions for $x_2 = 3$ and $0 \leq x_1 \leq 3$ are nondominated. In Fig. 2.2, these nondominated solutions are represented by the straight line BA. It is obvious that all the nondominated solutions must lie on the boundary BA of S_1 because any point interior to S_1 is dominated by at least one point on the boundary BA.

The nondominated solution concept is well utilized for the initial screen process in MADM. But the generation of a large number of nondominated solutions lessens significantly its effect of screening the feasible solutions in MODM; rather the nondominated concept is used for the sufficient condition of the final solution [44].

Satisficing solutions: A satisficing solution of Simon [373] is a reduced subset of the feasible set which exceeds all of the aspiration levels of each attribute. A set of satisficing solutions is composed of acceptable alternatives. Satisficing solutions need not be nondominated. This solution is credited for its simplicity which matches the behavior process of the DM whose knowledge and ability are limited-bounded rationality.

A satisficing solution may well be used as the final solution though it is often utilized for screening out unacceptable solutions.

A preferred solution: A preferred solution is a nondominated solution selected as the final choice through the DM's involvement in the information processing. In this respect the MCDM can be referred to as the decision aids to reach the preferred solution through utilizing the DM's preference information.

2. MODELS FOR MADM

A MADM method is a procedure that specifies how attribute information is to be processed in order to arrive at a choice. There are two major approaches in attribute information processing: noncompensatory and compensatory models. The two approaches are compared with the relevant MADM moethds.

2.1 Noncompensatory Model

These models do not permit tradeoffs between attributes. A disadvantage or unfavorable value in one attribute cannot be offset by an advantage or favorable value in some other attribute. Each attribute must stand on its own.

Hence comparisons are made on an attribute-by-attribute basis. The MADM methods which belong to this model are credited for their simplicity which matches the behavior process of the DM whose knowledge and ability are limited. They are dominance (1.1.1), maximin (1.1.2), maximax (1.1.3), conjunctive constraint method (2.1.1), disjunctive constraint method (2.1.2), and lexicographic method (2.2.1).

2.2. Compensatory Model

Compensatory models permit tradeoffs between attributes. That is, in these models changes (perhaps small changes only) in one attribute can be offset by opposing changes in any other attributes. With compensatory models a single number is usually assigned to each multidimensional characterization representing an alternative. Based upon the principle of calculating this number, compensatory models can be divided into three subgroups:

Scoring model: This model selects an alternative which has the highest score (or the maximum utility); therefore the problem is how to assess the appropriate multiattribute utility function for the relevant decision situation. Simple additive weighting (2.3.2), hierarchical additive weighting (2.3.3) and interactive simple additive weighting (3.1.2) belong to this model.

Compromising model: This model selects an alternative which is closest to the ideal solution. TOPSIS (2.3.5), LINMAP (3.1.1) and nonmetric MDS (3.2.1) belong to this category. Especially when a DM uses a square utility function, identification of an ideal solution is assisted by LINMAP procedures.

Concordance model: This model arranges a set of preference rankings which best satisfies a given concordance measure. Permutation method (2.2.3), linear assignment method (2.3.1), and ELECTRE (2.3.4) are considered to be in this model.

3. TRANSFORMATION OF ATTRIBUTES

An alternative in MADM is usually prescribed by two kinds of attributes: quantitative and qualitative (or fuzzy). In the example of the fighter aircraft problem, maximum speed, ferry range, maximum payload and cost are expressed in numerical or quantitative terms (in different units), but reliability and maneuverability in nonnumerical or qualitative terms. A question that arises is how to compare the two kinds of attributes. Further, how can we treat the nonhomogeneous units of measure (Mach, NM, lb, $, etc.)? These are the scaling problems. We are dealing with scaling problems without the concept of utility function.

There are three kinds of scales of measurement that can be employed for the measurement of quantities; ordinal, interval, and ratio [386, 405]. An ordinal scale puts measured entities (i.e., alternatives) in rank order but tells nothing of the relative distance between ranks. An interval scale provides equal intervals between entities and indicates the difference or distances of entities from some arbitrary origin. One outstanding example is that of the two common temperature scales (notice that the zero point is different for Fahrenheit and Centigrade scales). The ratio scale provides equal intervals between entities and indicates the difference or distances from some nonarbitrary origin. Ratio scales include weight (mg, lb, etc.), volumes (cc, cu ft, etc.), cost (cent, dollar, etc.), and so forth.

Since the transformation of a qualitative attribute into a ratio scale is extremely hard, most of the MADM methods resort to either the ordinal or the interval scale. The transformation of a qualitative attribute into the ordinal scale is much easier than into the interval scale. We will discuss the transformation of fuzzy attributes into interval scales first, then consider the problem of nonhomogeneous units of attributes after the quantification of fuzzy attributes.

3.1 Quantification of Fuzzy Attributes

One of the common ways for conversion of a qualitative attribute into an interval scale is to utilize the Bipolar scale [273]. For example, we may choose a 10-point scale and calibrate it in one of several ways. We can start with end points, giving 10 points to the maximum value that is practically or physically realizable and 0 points to the minimum attribute value that is practically or physically realizable. The midpoint would also be a basis for calibration, since it would be the breakpoint between values that are favorable (or better than average) and values that are unfavorable (or worse than average) [273].

Consider the attribute of reliability in the fighter aircraft decision problem: how many points should we assign to the value 'high'? Any value on the scale between 5.1 and 10.0 will satisfy the high rating; it will be on the scale and greater than the value assigned to 'average'. Commonly, values close to 10.0 will be reserved for extremely favorable characteristics; thus, for instance, 'very high' might be assigned the value 9.0. This in turn constrains 'high' to the interval 5.1-8.9. We might assign it the scale value 7.0. On the low end of the scale, 'very low' might be associated with the value 1.0 and 'low' with the value 3.0. These scale values are diagrammed in Fig. 2.4.

The procedures that derive these numerical values use addition and multiplication operations across attributes; from this, we note several implications. This type of scaling assumes that a scale value of 9.0 is

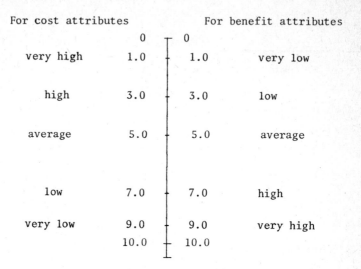

Fig. 2.4   Assignment of values for an interval scale.

three times as favorable as a scale value of 3.0. In addition, it assumes
that the difference between 'high' and 'low' is the same as the difference
between 'very low' and 'average' (4 scale points). Further, the combination
of values across attributes implies that the difference between any two
specific values (say, 'high' and 'low') is the same for each attribute.

It should be obvious that a numerical assignment such as that given
above is highly arbitrary. Many other scales are possible, for example
a scale of (very bad, bad, poor, fair, good, very good, excellent).
Sometimes attempts are made to provide some consistency checks [108].
Such checks are desirable but make the scaling procedure a very involved
activity posing many hypothetical questions to the DM.

3.2 Normalization

A normalization of the attribute values is not always necessary, but
it may be essential for some methods, like maximin, simple additive weighting,
ELECTRE, etc., to facilitate the computational problems inherent to the
presence of the different units in the decision matrix. A normalization
aims at obtaining comparable scales. There are different ways of normalizing
the attribute values.

Vector normalization: This procedure implies that each row vector
of the decision matrix is divided by its norm [305, BM-18], so that each
normalized value r_{ij} of the normalized decision matrix R can be calculated as

$$r_{ij} = \frac{x_{ij}}{\sqrt{\sum_{i=1}^{m} x_{ij}^2}} \tag{2.1}$$

This implies that all columns (attributes) have the same unit length of vector.

The advantage of this normalization is that all criteria are measured in dimensionless units, thus facilitating interattribute comparisons. A drawback is the fact that this normalization procedure does not lead to measurement scales with an equal length. The minimum and the maximum values of the scale are not equal to each criterion, so that a straight forward comparison is still difficult (due to the nonlinear scale trans- formation) [BM-18]. This procedure is utilized in the ELECTRE method and TOPSIS.

Linear scale transformation: A simple procedure is to divide the outcome of a certain criterion by its maximum value, provided that the criteria are defined as benefit criteria (the larger x_j, the greater pre- ference); then the transformed outcome of x_{ij} is

$$r_{ij} = \frac{x_{ij}}{x_j^*} \tag{2.2}$$

where $x_j^* = \max_i x_{ij}$. It is clear $0 \leq r_{ij} \leq 1$, and the outcome is more favorable as r_{ij} approaches 1. The advantage of this scale transformation is that all outcomes are transformed in a linear (proportional) way, so that the relative order of magnitude of the outcomes remains equal [BM-18]. In case of a cost criterion, r_{ij} has to be computed as

$$r_{ij} = 1 - \frac{x_{ij}}{x_j^*} \tag{2.3}$$

When both benefit and cost criteria exist in the decision matrix we should not use eqs. (2.2) and (2.3) at the same time because their bases are different, 0 for benefit criterion, 1 for cost. We can treat cost criteria as benefit criteria by taking the inverse of the outcomes (i.e., $1/x_{ij}$), and vice versa. Then eq. (2.2) for the cost criteria becomes

$$r_{ij} = \frac{1/x_{ij}}{\max_i (1/x_{ij})} = \frac{\min_i x_{ij}}{x_{ij}} = \frac{x_j^{\min}}{x_{ij}} \qquad (2.4)$$

When the number of benefit criteria is greater than that of the cost criteria, we treat all attributes as benefit criteria and use eqs. (2.2) and (2.4) simultaneously.

A more complicated form of r_{ij} with a benefit criterion may be

$$r_{ij} = \frac{x_{ij} - x_j^{\min}}{x_j^* - x_j^{\min}} \qquad (2.5)$$

In case of a cost criterion, r_{ij} is

$$r_{ij} = \frac{x_j^* - x_{ij}}{x_j^* - x_j^{\min}} \qquad (2.6)$$

The advantages of the definition of r_{ij} by eqs. (2.5) and (2.6) are that the scale of measurement varies precisely from 0 to 1 for each criterion. The worst outcome of a certain criterion implies $r_{ij} = 0$, while the best outcome implies $r_{ij} = 1$. A possible drawback of this procedure is that this scale transformation does not lead to a proportional change in outcomes.

4. FUZZY DECISION RULES

We already have mentioned the difficulty of handling the qualitative (or fuzzy) attributes in the MADM environment. Fuzzy attributes result from fuzzy goals (objectives) or fuzzy constraints. Let us consider an example of hiring a new faculty member in the Industrial Engineering department. The department sets the following set of goals:

i)   The candidate should be young

ii)   the candidate should be educated in a famous college

iii)   the candidate should be able to communicate well

iv)   the candidate should be experienced in industry

In these goals the underlined term is responsible for the fuzziness; then the associated attributes are also expressed in a fuzzy way:

i)   young, not young, very young, more or less young, not very young, old, not old, not very young and not very old, etc.

ii)   famous, not famous, very famous, more or less famous, not very famous, etc.

iii) and iv)   low, not low, very low, more or less low, medium high, not high, very high, more or less high, not low and not high, etc.

In the previous section of Transformation of Attributes we reviewed the ad hoc methods for converting fuzzy attributes into nonfuzziness, but these conventional quantitative techniques are not well suited for dealing with decision problems involving fuzziness. As decision making becomes more involved in both humanistic and complex systems, fuzziness becomes a prevalent phenomena in describing these systems. The basis for this contention is what Zadeh [456] calls the 'Principle of Incomparability', which he informally defines by stating: "as the complexity of a system increases, our ability to make precise and yet significant statements about its behavior diminishes until a threshold is reached beyond which precision and significance (or relevance) become almost mutually exclusive characteristics."

Zadeh [452] laid the initial foundations of the fuzzy set theory with his paper 'Fuzzy Sets' in 1965; since then the development in both theory and applications have been remarkable. Gaines and Kohout [137] listed 1150 papers in their review on the fuzzy set, and 45 papers were classified for decision making. We have found very few papers discussing fuzzy sets for the MADM environment [8, 19, 34, 257, 436, 440].

In this section we will review some of the basic concepts of fuzzy sets and their operations as it applies to the MADM methods. The text of Kaufmann [217] is especially recommended for the fuzzy set theory study.

4.1 Definition of Fuzzy Set [19, 217, 452]

Let E be a set and A a subset of E:

$$A \subset E$$

We usually indicate that an element x of E is a member of A using the symbol ε:

$$x \in A$$

To indicate this membership we may also use another concept, a characteristic function $\mu_A(x)$, whose value indicates whether x is a member of A:

$$\mu_A(x) = 1, \quad \text{if } x \in A$$

$$\mu_A(x) = 0, \quad \text{if } x \notin A$$

Imagine now that this characteristic function may take any value whatsoever in the interval [0,1]. Thus, an element of x_i of E may not be a member of A ($\mu_A = 0$), could be a member of A a little (μ_A near 0), may more or less be a member of A (μ_A neither too near 0 nor too near 1), could be strongly a member of A (μ_A near 1), or finally might be a member of A (μ_A = 1). In this way the notion of membership takes on extension and leads the definition of the fuzzy set.

Fuzzy subset: Let E be a set, denumberable or not, and let x be an element of E. Then a fuzzy subset A of E is a set of ordered pairs

$$A = \{ (x, \mu_A(x)) \} \quad \forall x \ \varepsilon \ E$$

where $\mu_A(x)$ is the grade or degree of membership of x in A. Thus, if $\mu_A(x)$ takes its values in a set M, called the membership set, one may say that x takes its value in M through the function $\mu_A(x)$. This function will be called the membership function. When M = {0, 1}, the fuzzy subset A is nonfuzzy and its membership function becomes identical with the characteristic function of a nonfuzzy set. We may conclude that a fuzzy set is a class of objects in which there is no sharp boundary between two objects, one belonging to the class and one not.

Numerical Example

E = {Australia, Canada, China, France, India, Japan, Yugoslavia}

A_1 = the communist countries

   = {(Australia, 0.), (Canada, 0.), (China, 1.), (France, 0.) (India, 0.),

   (Japan, 0.), (Yugoslavia, 1.)}

   = {(China, 1.), (Yugoslavia, 1.)}

   = {(China, Yugoslavia)}

A_2 = the faithful allies to the U.S.A.

   = {(Australia, .8), (Canada, .95), (China, .05), (France, .8), (India, .3)

   (Japan, .7), (Yugoslavia, .01)}

Note that A_1 is a nonfuzzy subset and A_2 is a fuzzy subset of E. The values of $\mu_{A_2}(\cdot)$ are selected subjectively but it does give us a handle for dealing with subjective concepts in a rational way, in a similar manner as the method used by the Bayesian decision makers enables them to handle sub-

jective probabilities and utilities [437].

4.2 Some Basic Operations of Fuzzy Sets [19, 217, 452]

Equality: Two fuzzy sets are equal, written as A = B, if and only if $\mu_A = \mu_B$, that is

$$\mu_A(x) = \mu_B(x) \qquad \forall x \in E$$

Containment: A fuzzy set A is contained in or is a subset of a fuzzy set B, written as A ⊂ B, if and only if

$$\mu_A(x) \leq \mu_B(x) \qquad \forall x \in E$$

Complementation: B is said to be the complement of A if and only if

$$\mu_B(x) = 1 - \mu_A(x) \qquad \forall x \in E$$

Intersection: The intersection of A and B is denoted by A ∩ B and is defined as the largest fuzzy set contained in both A and B. The membership function of A ∩ B is given by

$$\mu_{A \cap B}(x) = Min(\mu_A(x), \mu_B(x)) \qquad \forall x \in E$$

Union: The union of A and B, denoted as A U B, is defined as the smallest set containing both A and B. The membership function of A U B is given by

$$\mu_{AUB}(x) = Max(\mu_A(x), \mu_B(x)) \qquad \forall x \in E$$

Numerical Example

Let E = {Amy, Becky, Cindy, Donna}

    A = a set of beautiful girls

      = {(Amy, .7), (Becky, .2), (Cindy, .8), (Donna, .6)}

    B = a set of intelligent girls

      = {(Amy, .5), (Becky, .9), (Cindy, .4), (Donna, .7)}

Then, the set of both beautiful and intelligent girls is:

A ∩ B = {(Amy, .5), (Becky, .2), (Cindy, .4), (Donna, .6)}

Also the set of beautiful or intelligent girls is:

A ∪ B = {(Amy, .7), (Becky, .9), (Cindy, .8), (Donna, .7)}

Algebraic product: The algebraic product of A and B is denoted by AB and is defined by

$$\mu_{AB}(x) = \mu_A(x) \cdot \mu_B(x) \qquad \forall x \in E$$

Thus A^α, where α is any positive number, has its membership function of

$$\mu_A(x)^\alpha \qquad \forall x \in E$$

If $\alpha > 1$, the effect of raising A to the α power results in a fuzzy subset A^α (i.e., $A^\alpha \subset A$) in which the degree of membership for those x that are large is reduced much less than for those that are small. If $\alpha < 1$, the result is the opposite ($A \subset A^\alpha$).

Numerical Example

Let A = {(x_1, .3), (x_2, .4), (x_3, .7), (x_4, .9)}

then

$$A^2 = \{(x_1, .09), (x_2, .16), (x_3, .49), (x_4, .81)\}$$
$$A^{\frac{1}{2}} = \{(x_1, .55), (x_2, .63), (x_3, .84), (x_4, .95)\}$$

Also we can see the relation of

$$A^2 \subset A \subset A^{\frac{1}{2}}$$

Algebraic sum: The algebraic sum of A and B is denoted by A ⊕ B, and is defined by

$$\mu_{A \oplus B}(x) = \mu_A(x) + \mu_B(x) - \mu_A(x)\mu_B(x) \qquad \forall x \in E$$

<u>Fuzzy set induced by mappings</u>: Let $f: E_1 \to E_2$ be a mapping from $E_1 = \{x\}$ to $E_2 = \{y\}$, with the image of x under f denoted by $y = f(x)$. Let A be a fuzzy set in E_1. Then, the mapping f induces a fuzzy set B in E_2 whose membership function is given by

$$\mu_B(y) = \underset{x \in f^{-1}(y)}{\text{Sup}} \mu_A(x)$$

where the supreme has taken over the set of points $f^{-1}(y)$ in E_1 which are mapped by f into y.

<u>Numerical Example</u>

Let

$E_1 = \{Amy, Becky, Cindy, Donna\}$

$E_2 = \{Bill, David, Tom\}$

The fuzzy subset A of E_1 is the set of beautiful girls.

A = {(Amy, .7), (Becky, .2), Cindy, .8), (Donna, .6)}

Assume that the following information is available:

Amy is willing to marry Bill   Cindy is willing to marry David

Becky is willing to marry Bill   Donna is willing to marry David or Tom

Then what is the fuzzy subset B of E_2, of which a boy has a beautiful wife?

First consider the inverse mapping f^{-1} of

$f^{-1}(Bill) = \{Amy, Becky\}$

$f^{-1}(David) = \{Cindy, Donna\}$

$f^{-1}(Tom) = \{Donna\}$

Then we can have

$$\mu_B(Bill) = \underset{x \in f^{-1}(Bill)}{\text{Sup}} \mu_A(x)$$

$$= \text{Sup}\ (.7, .2) = .7$$

$\mu_B(\text{David}) = \text{Sup}(.8, .6) = .8$

$\mu_B(\text{Tom}) = \text{Sup}(.6) = .6$

The resulting fuzzy subset B is

$$B = \{(\text{Bill}, .7), (\text{David}, .8), (\text{Tom}, .6)\}$$

The foregoing process is illustrated by Fig. 2.5.

Conditioned fuzzy sets: A fuzzy set B(x) in $E_2 = \{y\}$ is conditioned on x if its membership function depends on x as a parameter. This dependence is expressed by $\mu_B(y|x)$. Suppose that the parameter x ranges over a space E_1, so that to each x in E_1 corresponds a fuzzy set B(x) in E_2. Through this mapping, any given fuzzy set A in E_1 induces a fuzzy set B in E_2 which is defined by

$$\mu_B(y) = \underset{x \in E_1}{\text{Sup Min}} (\mu_B(y|x), \mu_A(x))$$

Numerical Examples

Assume that the reliability of the fighter aircraft depends upon the maintenance skill of a country. Let us consider the four countries which use three kinds of American-made fighter aircraft:

$E_1 = \{\text{Greece, Saudi Arabia, W. Germany, United Kingdom}\}$

$E_2 = \{\text{F-4E, F-15, F/A-18}\}$

The fuzzy set A of maintenance skill is given as

$A = \{(G, .8), (S, .5), (W, .9), (U, .95)\}$

The conditional fuzzy set (the maintainability of an aircraft with respect to a country) is

$\mu_B(F|G) = \{(\text{F-4E}, .85), (\text{F-15}, .75), (\text{F/A-18}, .7)\}$

$\mu_B(F|S) = \{(\text{F-4E}, .6), (\text{F-15}, .5), (\text{F/A-18}, .5)\}$

$\mu_B(F|W) = \{(\text{F-4E}, .95), (\text{F-15}, .85), (\text{F/A-18}, .8)\}$

$\mu_B(F|U) = \{(\text{F-4E}, .95), (\text{F-15}, .9), (\text{F/A-18}, .85)\}$

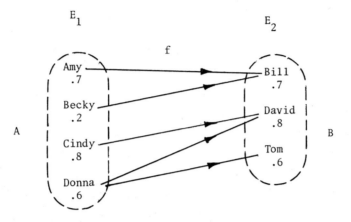

Fig. 2.5   Fuzzy subset B induced by a mapping.

Then $\mu_B(\text{F-4E})$ is:

Min $[\mu_B(\text{F-4E}|G), \mu_A(G)]$ = Min $(.85, .8)$ = $.8$

Min $[\mu_B(\text{F-4E}|S), \mu_A(S)]$ = Min $(.6, .5)$ = $.5$

Min $[\mu_B(\text{F-4E}|W), \mu_A(W)]$ = Min $(.95, .9)$ = $.9$

Min $[\mu_B(\text{F/A-18}|U), \mu_A(U)]$ = Min $(.95, .95)$ = $.95$

$\mu_B(\text{F-4E})$ = Sup Min $[\mu_B(\text{F-4E}|x_i), \mu_A|x_i)]$
$\qquad\quad x_i$

$\qquad$ = Sup $(.8, .5, .9, .95)$

$\qquad$ = $.95$.

Similary we obtain $\mu_B(\text{F-15})$ = $.9$, and $\mu_B(\text{F/A-18})$ = $.85$.

Thus the resulting fuzzy set B (the reliability of each fighter is)

$\quad$ B = {(F-4E, .95), (F-15, .9), (F/A-18, .85)}.

5. METHODS FOR ASSESSING WEIGHT

Many methods for MCDM problems require information about the relative importance of each attribute (or objective). It is usually given by a set of (preference) weights which is normalized to sum to 1. In case of n criteria, a set of weights is

$$\underline{w}^T = (w_1, w_2, \ldots, w_j, \ldots, w_n)$$

$$\sum_{j=1}^{n} w_j = 1$$

Eckenrode [85] suggests six techniques for collecting the judgments of decision makers concerning the relative value of criteria. They are simple, but to a single DM, we need more elegant methods to substitute a small (single) statistical sample for a large one. We will deal with four techniques recently developed: eigenvector method, weighted least square method, entropy method, and LINMAP. Entropy method and LINMAP can not be used in MODM environment because they need decision matrix as a part of input.

5.1 Eigenvector Method

The DM is supposed to judge the relative importance of two criteria. The number of judgments is $_nC_2 = n(n-1)/2$. Some inconsistencies from these judgments are allowed. Saaty [350] introduced a method of scaling ratios using the principle eigenvector of a positive pairwise comparison matrix.

Let matrix A be

$$A = \begin{bmatrix} a_{11} & a_{12} & \cdots & a_{1n} \\ a_{21} & a_{22} & \cdots & a_{2n} \\ \cdot & \cdot & & \cdot \\ \cdot & \cdot & & \cdot \\ \cdot & \cdot & & \cdot \\ a_{n1} & a_{n2} & \cdots & a_{nn} \end{bmatrix}$$

$$
= \begin{bmatrix}
\dfrac{w_1}{w_1} & \dfrac{w_1}{w_2} & \cdots & \dfrac{w_1}{w_n} \\[2ex]
\dfrac{w_2}{w_1} & \dfrac{w_2}{w_2} & \cdots & \dfrac{w_2}{w_n} \\[2ex]
\cdot & \cdot & & \cdot \\
\cdot & \cdot & & \cdot \\
\cdot & \cdot & & \cdot \\[1ex]
\dfrac{w_n}{w_1} & \dfrac{w_n}{w_2} & & \dfrac{w_n}{w_n}
\end{bmatrix}
\tag{2.7}
$$

This is a 'reciprocal matrix' which has all positive elements and has the reciprocal property

$$
a_{ij} = 1/a_{ji} \tag{2.8}
$$

and

$$
a_{ij} = a_{ik}/a_{jk} \tag{2.9}
$$

Multiplying A by $\underline{w} = (w_1, w_2, \ldots, w_n)^T$ yields

$$
A \underline{w} =
\begin{bmatrix}
\dfrac{w_1}{w_1} & \dfrac{w_1}{w_2} & \cdots & \dfrac{w_1}{w_n} \\[2ex]
\dfrac{w_2}{w_1} & \dfrac{w_2}{w_2} & \cdots & \dfrac{w_2}{w_n} \\[2ex]
\cdot & \cdot & & \cdot \\
\cdot & \cdot & & \cdot \\
\cdot & \cdot & & \cdot \\[1ex]
\dfrac{w_n}{w_1} & \dfrac{w_n}{w_2} & \cdots & \dfrac{w_n}{w_n}
\end{bmatrix}
\begin{bmatrix}
w_1 \\[1ex] w_2 \\ \cdot \\ \cdot \\ \cdot \\ w_n
\end{bmatrix}
= n
\begin{bmatrix}
w_1 \\[1ex] w_2 \\ \cdot \\ \cdot \\ \cdot \\ w_n
\end{bmatrix}
= n \, \underline{w}
$$

or

$$
(A - nI) \, \underline{w} = 0 \tag{2.10}
$$

Due to the consistency property of eq. (2.9), the system of homogeneous linear equations, eq.(2.10) has only trivial solutions.

In general, the precise values of w_i/w_j are unknown and must be estimated. In other words, human judgments can not be so accurate that eq. (2.9) be satisfied completely. We know that in any matrix, small perturbations in the coefficients imply small perturbations in the eigenvalues. If we define A´ as the DM's estimate of A and $\underline{w}´$ is corresponding to A´, then

$$A´\underline{w}´ = \lambda_{max} \underline{w}´ \tag{2.11}$$

where λ_{max} is the largest eigenvalue of A´. $\underline{w}´$ can be obtained by solving the system of linear equations, eq. (2.11)

Numerical Example

If the following positive pairwise comparison matrix is given

$$A = \begin{bmatrix} 1 & 1/3 & 1/2 \\ 3 & 1 & 3 \\ 2 & 1/3 & 1 \end{bmatrix}$$

then set the determinant of (A - λ I) as zero. That is

$$\det (A - \lambda I) = \begin{vmatrix} 1-\lambda & 1/3 & 1/2 \\ 3 & 1-\lambda & 3 \\ 2 & 1/3 & 1-\lambda \end{vmatrix} = 0$$

The largest eigenvalue of A, λ_{max}, is 3.0536, and we have

$$\begin{bmatrix} -2.0536 & \frac{1}{3} & \frac{1}{2} \\ 3 & -2.0536 & 3 \\ 2 & \frac{1}{3} & -2.0536 \end{bmatrix} \begin{bmatrix} w_1 \\ w_2 \\ w_3 \end{bmatrix} = 0$$

The solution of the homogeneous system of linear equations gives (recall that $\sum\limits_{i=1}^{3} w_i = 1$)

$$\underline{w}^T = (0.1571, 0.5936, 0.2493).$$

The Scale

For assessing the scale ratio w_i/w_j, Saaty [350] gives an intensity scale of importance for activities and has broken down the importance ranks as shown in Table 2.2.

Numerical Example (The Wealth of Nations through their World Influence [350])

It is assumed that the world influence of nations is a function of six factors: (1) human resources; (2) wealth; (3), trade; (4) technology; (5) military power; and (6) potential natural resources (such as oil).

Seven countries (the United States, the U.S.S.R., China, France, the United Kingdom, Japan, and West Germany) as a group comprise a dominant class of influential nations. It is desired to compare them among themselves as to their overall influence in international relations. For a very rough estimate and for illustration of the method, we consider only the single factor of wealth. The question to answer is: how much more strongly does one nation as compared with another contribute its wealth to gain world influence?

The first row of Table 2.3 gives the pairwise comparison of the wealth contributed by the United States with wealth contributed by the other nations. For example, it is of equal importance to the United States (hence, the unit entry), between weak importance and strong importance when compared with the U.S.S.R. (hence, the value of 4), of absolute importance when compared with China (hence, the value of 9), between strong and demonstrated importance when compared with France and the United Kingdom (hence, 6 for both), strong importance when compared with Japan and West Germany (hence, 5 for both). The reciprocals, of the numbers in the first row are entered into the entries in the first column, which indicates the inverse relation of relative strength of the wealth of the other countries when compared with the United States, and so on, for the remaining values in the second row and second column, etc.

Table 2.2   The scale and its description [350]

Intensity of importance	Definition	Explanation
1	Equal importance	Two criteria contribute equally to the objective.
3	Weak importance of one over another	Experience and judgment slightly favor one criterion over another.
5	Essential or strong importance	Experience and judgment strongly favor one criterion over another.
7	Demonstrated importance	A criterion is strongly favored and its dominance is demonstrated in practice.
9	Absolute importance	The evidence favoring one criterion over another is of the highest possible order of affirmation.
2,4,6,8	Intermediate values between the two adjacent judgments	When compromise is needed.

Table 2.3 Wealth comparison matrix

Country	US	USSR	China	France	UK	Japan	W. Germany
US	1	4	9	6	6	5	5
USSR	1/4	1	7	5	5	3	4
China	1/9	1/7	1	1/5	1/5	1/7	1/5
France	1/6	1/5	5	1	1	1/3	1/3
UK	1/6	1/5	5	1	1	1/3	1/3
Japan	1/5	1/3	7	3	3	1	2
W. Germany	1/5	1/4	5	3	3	1/2	1

Table 2.4 Normalized wealth eigenvector

Country	Normalized eigenvector	Actual GNP$^\alpha$(1972)	Fraction of GNP Total
U.S.	0.429	1167	0.413
USSR	0.231	635	0.225
China	0.021	120	0.043
France	0.053	196	0.069
U.K.	0.053	154	0.055
Japan	0.119	294	0.104
W. Germany	0.095	257	0.091
Total		2823	

$^\alpha$Billions of dollars

The largest eigenvalue of the wealth matrix given by Table 2.3 is 7.61 and the normalized eigenvector is presented in Table 2.4

It is interesting to note that the comparisons in the wealth matrix are not consistent. For example, U.S. vs U.S.S.R. = 4, U.S.S.R. vs China = 7, but U.S. vs China = 9, not 28. Nevertheless, the relative weights of .429 and .231 for the United States and Russia are obtained, and these weights are in striking agreement with the corresponding GNP's as percentages of the total GNP as shown in Table 2.4. Despite the apparent arbitrariness of the scale, the irregularities disappear and the numbers occur in good accord with observed data.

Note

Johnson [212] indicates the right-left asymmetry in an eigenvector procedure. Essentially, by analogy to eq. (2.11) it is suggested that the prioritization be constructed in the same way when A is not consistent. However, for a left eigenvector $\underline{w}^{(L)}$,

$$\underline{w}^{(L)T} A = \lambda_{max} \underline{w}^{(L)T} \qquad (2.12)$$

In the consistent case there is a constant c such that

$$w_j^{(L)} = c \ (1/w_j), \qquad j = 1,2,\ldots,m$$

That is, the componentwise inverse of a right eigenvector is a left eigenvector. However, for $m \geq 4$ the two can disagree even when matrix A is consistent. Johnson et al. note that use of the left eigenvectors is equally justified (as long as order is reversed).

5.2 Weighted Least Square Method

A weighted least square method is proposed by Chu et al. [54] to obtain the weight. This method has the advantage that it involves the solution of a set of simultaneous linear algebraic equations and is conceptually easier to understand than Saaty's eigenvector method [350].

Consider the elements a_{ij} of Saaty's matrix A in eq. (2.7). It is desired to determine the weights w_i, such that, given a_{ij}

$$a_{ij} \simeq w_i/w_j \tag{2.13}$$

The weights can be obtained by solving the constrained optimization problem

$$\min \; z = \sum_{i=1}^{n} \sum_{j=1}^{n} (a_{ij}w_j - w_i)^2 \tag{2.14}$$

$$\text{s.t.} \quad \sum_{i=1}^{n} w_i = 1 \tag{2.15}$$

An additional constraint is that $w_i > 0$. However, it is conjectured that the above problem can be solved to obtain $w_i > 0$ without this constraint.

In order to minimize z, the Lagrangian function is formed.

$$L = \sum_{i=1}^{n} \sum_{j=1}^{n} (a_{ij}w_j - w_i)^2 + 2\lambda(\sum_{i=1}^{n} w_i - 1) \tag{2.16}$$

where λ is the Lagrangian multiplier. Differentiating eq. (2.16) with respect to w_ℓ, the following set of equations is obtained:

$$\sum_{i=1}^{n} (a_{i\ell}w_\ell - w_i) a_{i\ell} - \sum_{j=1}^{n} (a_{\ell j}w_j - w_\ell) + \lambda = 0, \quad \ell = 1,2,\ldots,n \tag{2.17}$$

Eqs. (2.17) and (2.15) form a set of (n + 1) nonhomogeneous linear equations with (n + 1) unknowns. For example, for n = 2, the equations are (recall that $a_{ii} = 1$ ∀i):

$$(1 + a_{21}^2) w_1 - (a_{12} + a_{21}) w_2 + \lambda = 0$$

$$- (a_{21} + a_{12}) w_1 + (1 + a_{12}^2) w_2 + \lambda = 0$$

$$w_1 + w_2 = 1$$

Given the coefficients a_{ij}, the above equations can be solved for w_1, w_2, and λ.

In general, eqs. (2.17) and (2.15) can be expressed in the matrix form

$$B\underline{w} = \underline{m} \qquad\qquad (2.18)$$

where

$$\underline{w} = (w_1, w_2, \ldots, w_n, \lambda)^T$$

$$\underline{m} = (0, 0, \ldots, 0, 1)^T$$

B = (m + 1) x (n + 1) matrix with elements b_{ij}

$$b_{ii} = (n - 1) + \sum_{j=i}^{n} a_{ji}^2, \qquad i, j = 1, \ldots, n$$

$$b_{ij} = -(a_{ij} + a_{ji}), \qquad i, j = 1, \ldots, n$$

$$b_{k,n+1} = b_{n+1,k} = 1, \qquad k = 1, \ldots, n$$

$$b_{n+1,n+1} = 0$$

Numerical Example 1.

Given Saaty's matrix A of pairwise comparisons:

$$A = \begin{bmatrix} 1 & 1/3 & 1/2 \\ 3 & 1 & 3 \\ 2 & 1/3 & 1 \end{bmatrix}$$

For n = 3, eq. (2.18) becomes:

$$(a_{21}^2 + a_{31}^2 + 2)w_1 - (a_{12} + a_{21})w_2 - (a_{13} + a_{31})w_3 + \lambda = 0$$

$$-(a_{21} + a_{12})w_1 + (a_{12}^2 + a_{32}^2 + 2)w_2 - (a_{23} + a_{32})w_3 + \lambda = 0$$

$$-(a_{31} + a_{13})w_1 - (a_{32} + a_{23})w_2 + (a_{13}^2 + a_{23}^2 + 2)w_3 + \lambda = 0$$

$$w_1 + w_2 + w_3 = 1$$

Substituting values for a_{ij} into the above equations, we obtain

$$15w_1 - \frac{10}{3} w_2 - \frac{5}{2} w_3 + \lambda = 0$$

$$- \frac{10}{3} w_1 + \frac{20}{9} w_2 - \frac{10}{3} w_3 + \lambda = 0$$

$$- \frac{5}{2} w_1 - \frac{10}{3} w_2 + \frac{45}{4} w_3 + \lambda = 0$$

$$w_1 + w_2 + w_3 = 1$$

The solution is $\underline{w}^T$ = (0.1735, 0.6059, 0.2206), which is reasonably close to the solution, $\underline{w}^T$ = (0.1571, 0.5936, 0.2493), obtained by Saaty's eigenvector method.

Numerical Example 2. (The Wealth of Nations through their World Influence)

The weighted least square method is applied to Saaty's wealth comparison matrix (See Table 2.3). The results are compared with Saaty's results as presented in Table 2.5 [54].

Table 2.5   Comparison of numerical results for wealth comparison matrix

Country	Saaty's eigenvector (λ_{max} = 7.61)	Weighted least-square method
US	0.429	0.4867
USSR	0.231	0.1750
China	0.021	0.0299
France	0.053	0.0593
UK	0.053	0.0593
Japan	0.119	0.1043
W. Germany	0.095	0.0853

5.3 Entropy Method

When the data of the decision matrix is known, instead of the Saaty's pairwise comparison matrix as given by eq. (2.7), the entropy method and the LINMAP method can be used for evaluating the weights.

Entropy has become an important concept in the social sciences [46, 305] as well as in the physical sciences. In addition entropy has a useful meaning in information theory, where it measures the expected information content of a certain message. Entropy in information theory is a criterion for the amount of uncertainty represented by a discrete probability distribution, p_i, which agrees that a broad distribution represents more uncertainty than does a sharply peaked one [209]. This measure of uncertainty is given by Shannon [365] as

$$S(p_1, p_2, \ldots, p_n) = -k \sum_{j=1}^{n} p_j \ln p_j \qquad (2.19)$$

where k is a positive constant. Since this is just the expression for entropy as found in statistical mechanics, it is called the entropy of the probability distribution p_i; hence the terms "entropy" and "uncertainty" are considered as synonymous. When all p_i are equal to each other for a given i, $p_i = 1/n$, $S(p_1, \ldots, p_n)$ takes on its maximum value.

The decision matrix for a set of alternatives contains a certain amount of information, entropy can be used as a tool in criteria evaluation [305 459]. The entropy idea is particularly useful to investigate contrasts between sets of data. For example, a criterion does not function much when all the alternatives have the similar outcomes for that criterion. Further, if all the values are the same, we can eliminate the criterion.

Since a project outcome p_{ij} includes a certain information content, the information content of the project outcomes of criteria j can be measured by means of the entropy value. However, the significance of p_{ij} is determined by the differential outcomes of all alternatives, so that p_{ij} may be adapted for the average intrinsic information generated by the set of alternatives through criterion j.

Let the decision matrix D of m alternatives and n attributes (criteria) be

$$
D = \begin{array}{c} \\ A_1 \\ A_2 \\ \cdot \\ \cdot \\ \cdot \\ A_m \end{array}
\begin{array}{cccc} X_1 & X_2 & \cdots & X_n \end{array}
\begin{bmatrix} x_{11} & x_{12} & \cdots & x_{1n} \\ x_{21} & x_{22} & & x_{2n} \\ \cdot & \cdot & & \\ \cdot & \cdot & & \\ \cdot & \cdot & & \\ x_{m1} & x_{m2} & \cdots & x_{mn} \end{bmatrix}
$$

The project outcomes of attribute j, p_{ij}, then can be defined as

$$
p_{ij} = \frac{x_{ij}}{\sum\limits_{i=1}^{m} x_{ij}}, \quad \forall i,j \tag{2.20}
$$

The entropy E_j of the set of project outcomes of attribute j is:

$$
E_j = -k \sum_{i=1}^{m} p_{ij} \ln p_{ij}, \quad \forall j \tag{2.21}
$$

where k represents a constant:

$$
k = 1/\ln m
$$

which guarantees that $0 \leq E_j \leq 1$.

The degree of diversification d_j of the information provided by the outcomes of attribute j can be defined as

$$d_j = 1 - E_j, \quad \forall j \qquad (2.22)$$

If the DM has no reason to prefer one criterion over another, the Principle of Insufficient Reason [384] suggests that each one should be equally preferred. Then the best weight set he can expect, instead of the equal weight, is

$$w_j = \frac{d_j}{\sum\limits_{j=1}^{n} d_j}, \quad \forall j \qquad (2.23)$$

If the DM has a prior, subjective weight λ_j, then this can be adapted with the help of w_j information. The new weight w_j^{0} is,

$$w_j^{0} = \frac{\lambda_j w_j}{\sum\limits_{j=1}^{n} \lambda_j w_j}, \quad \forall j \qquad (2.24)$$

Numerical Example (A Fighter Aircraft Selection Problem)

Let us consider the following decision matrix of 4 jet fighters with 6 attributes

$$
D = \begin{array}{c} \\ A_1 \\ A_2 \\ A_3 \\ A_4 \end{array}
\begin{array}{cccccc}
X_1 & X_2 & X_3 & X_4 & X_5 & X_6 \\
\left[\begin{array}{cccccc}
2.0 & 1500 & 20,000 & 5.5 & 5 & 9 \\
2.5 & 2700 & 18,000 & 6.5 & 3 & 5 \\
1.8 & 2000 & 21,000 & 4.5 & 7 & 7 \\
2.2 & 1800 & 20,000 & 5.0 & 5 & 5
\end{array}\right]
\end{array}
$$

By using eq. (2.20) we obtain p_{ij} for all i and j:

$$
[p_{ij}] = \begin{array}{c} \\ A_1 \\ A_2 \\ A_3 \\ A_4 \end{array}
\begin{array}{cccccc}
X_1 & X_2 & X_3 & X_4 & X_5 & X_6 \\
\begin{bmatrix} .2353 & .1875 & .2532 & .2558 & .25 & .3462 \\
.2941 & .3375 & .2278 & .3023 & .15 & .1923 \\
.2118 & .2500 & .2658 & .2093 & .35 & .2692 \\
.2588 & .2250 & .2532 & .2326 & .25 & .1923 \end{bmatrix}
\end{array}
$$

The entropy of each attribute, E_j, the degree of diversification, d_j, and the normalized weight, w_j, are calculated by using eqs. (2.21), (2.22), and (2.23), respectively. They are:

	X_1	X_2	X_3	X_4	X_5	X_6
E_j	.9946	.9829	.9989	.9931	.9703	.9770
d_j	.0054	.0171	.0011	.0069	.0297	.0230
w_j	.0649	.2055	.0133	.0829	.3570	.2764

The results indicate that the weight of importance is in the order of $(w_5 = .3570,\ w_6 = .2764,\ w_2 = .2055,\ w_4 = .0829,\ w_1 = .0649,\ w_3 = .0133)$. The weight of an attribute is smaller when all the alternatives of the attribute have similar outcomes.

If the DM has the following a priori weight λ_j:

$$\underline{\lambda} = (.2, .1, .1, .1, .2, .3)$$

Then the subjective weights which are calculated by using eq. (2.24) are

$$\underline{w}^0 = (.0657, .1041, .0067, .0420, .3616, .4199)$$

Note

Zeleny [459] uses another value of p_{ij} in eq. (2.20). He first measures the closeness to the ideal solution, r_{ij}, such as

$$r_{ij} = \frac{x_{ij}}{x_j^*}$$

or

$$r_{ij} = \frac{x_{ij} - x_j^{min}}{x_j^* - x_j^{min}}$$

where $x_j^* = \max_i x_{ij}, \quad \forall j$

$x_j^{min} = \min_i x_{ij}, \quad \forall_j$

then p_{ij} is defined as

$$p_{ij} = \frac{r_{ij}}{\sum\limits_{i=1}^{m} r_{ij}}, \quad \forall i, j \qquad (2.25)$$

The weights of importance obtained by using eq. (2.25), however, are very close to those obtained by using eq. (2.20).

5.4 LINMAP

Srinivasan and Shocker [380] developed LINMAP (LINear programming techniques for Multidimensional Analysis of Preference) for assessing the weights of attributes as well as for selecting the alternative. In this method m alternatives composed of n attributes are represented as m points in the n-dimensional space. A DM is assumed to have his ideal point denoting his most preferred stimulus (alternative) location in this space and a set of weights which reveal the relative saliences of the attributes. He prefers those stimuli which are "closer" to his ideal point (in terms of a weighted Euclidean distance measure). A linear programming model is proposed for "external analysis" i.e., estimation of the coordinates of his ideal point and the weights (involved in the Euclidean distance measure) by analyzing his paired comparison preference judgments on a set of stimuli, prespecified by their coordinate locations in the multidimensional space.

The details of the method will be presented in the section 3.1.1 LINMAP of Part IV Methods for Multiple Attribute Decision Making.

III. METHODS FOR MULTIPLE ATTRIBUTE DECISION MAKING

1. METHODS FOR NO PREFERENCE INFORMATION GIVEN

There are some classical decision rules such as dominance, maximin and maximum which are still fit for the MADM environment. They do not require the DM's preference information, and accordingly yield the objective (vs. subjective) solution. However, the right selection of these methods for the right situation is important. (See Table 1.3 for references).

1.1.1 DOMINANCE

An alternative is dominated if there is another alternative which excels it in one or more attributes and equals it in the remainder. The number of alternatives can be reduced by eliminating the dominated ones. In other words we screen the set of alternatives before the final choice is made. A set of nondominated solutions is one obtained through the sieve of dominance method.

This method does not require any assumption or any transformation of attributes. The sieve of dominance takes the following procedures [44]; compare the first two alternatives and if one is dominated by the other, discard the dominated one. Next compare the undiscarded alternatives with the third alternative and discard any dominated alternative. Then introduce the fourth alternative and so on. After (m - 1) stages the nondominated set is determined. This nondominated set usually has multiple elements in it, hence the dominance method is mainly used for the initial filtering [400].

Numerical Example (The Fighter Aircraft Problem)

The decision matrix of the fighter aircraft problem is

$$
D = \begin{array}{cccccc}
 X_1 & X_2 & X_3 & X_4 & X_5 & X_6 \\
\end{array}
$$

	X_1	X_2	X_3	X_4	X_5	X_6	
	2.0	1500	20,000	5.5	average	very high	A_1
$D =$	2.5	2700	18,000	6.5	low	average	A_2
	1.8	2000	21,000	4.5	high	high	A_3
	2.2	1800	20,000	5.0	average	average	A_4

Recall that X_4 is a cost criterion. In this example, all alternatives are nondominated.

If A_1 and A_4 had the same rate of maneuverability (X_6), then A_4 would dominate A_1 because X_1, X_2 and X_4 of A_4 excel those of A_1, and other attributes are equal.

NOTE

Calpine and Golding [44] derived the expected number of nondominated solutions when m alternatives are compared with respect to n attributes. Consider first the very special case in which all the elements in the decision matrix are random numbers uniformly distributed over the range 0 to M. Attention is first focussed on the final (n^{th}) column. Arrange the rows so that the elements in the nth column are in decreasing order of magnitude. By the randomness of the elements, the probability of an arbitrarily selected row being the rth in the order is $1/m$. Let $p(m, n)$ be the probability that a row, arbitrarily chosen from m rows (alternatives), is nondominated with respect to n attributes. Consider the rth row. The ordering ensures that this row is not dominated by any row below it and also that it exceeds no row above it in the nth attribute. Hence a necessary and sufficient conditions for the rth row to be nondominated is that it is nondominated among the first r candidates with respect to the first $(n - 1)$ attributes. Thus the probability of a row being the rth and nondominated is $p(r, n-1)/m$. The probability of an arbitrarily selected row being nondominated is

$$p(m, n) = \sum_{r=1}^{m} p(r, n - 1)/m$$

$$= \{p(m, n - 1) + (m - 1) p(m - 1, n)\}/m \qquad (3.1)$$

Then the average number of nondominated alternatives, $a(m, n)$, is

$$a(m, n) = mp(m, n)$$

$$= a(m, m - 1)/m + a(m - 1, n)$$

As $a(m, 1) = a(1, n) = 1$, $a(m, n)$ can be calculated recursively. A good approximation of $a(m, n)$ is given by

$$a(m, n) \approx 1 + \ln m + (\ln m)^2/2! + \cdots$$

$$+ (\ln m)^{n-3}/(n - 3)! + \gamma(\ln m)^{n-2}/(n - 2)!$$

$$+ (\ln m)^{n-1}/(n - 1)! \qquad (3.2)$$

where γ is Euler's constant (~ 0.5772). Some typical results are shown in Fig. 3.1.

It indicates that the number of nondominated alternatives, for a few attributes, e.g., n=4, will be reduced to 8, 20, and 80 for m=10, 100, and 1000, respectively; however, the number of nondominated alternatives for a large number of attributes, e.g., n=8, will still be very large, i.e., 10, 90, and 900, respectively for m=10, 100, and 1000.

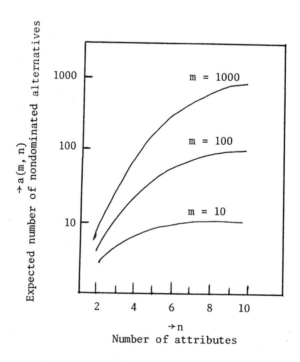

Fig. 3.1. Expected number of nondominated alternatives [44].

1.1.2   MAXIMIN

An astronaut's life or death in the orbit may depend upon his worst vital organ, and a chain is only as strong as its weakest link. In this situation where the overall performance of an alternative is determined by the weakest or poorest attribute, a DM would examine the attribute values for each alternative, note the lowest value for each alternative, and then select the alternative with the most acceptable value in its lowest attribute. It is the selection of the maximum (across alternatives) of the minimum (across attributes) values, or the maximin.

Under this procedure only a single weakest attribute represents an alternative; all other (n-1) attributes for a particular alternative are ignored. If these lowest attribute values come from different attributes, as they often do, we may be basing our final choice on single values of attributes that differ from alternative to alternative. Therefore, the maximin method can be used only when interattribute values are comparable; that is, all attributes must be measured on a common scale; however, they need not be numerical. The alternative, A^+, is selected such that

$$A^+ = \{A_i \mid \max_i \min_j x_{ij} \}, \quad j = 1, 2, \ldots, n; \quad i = 1, 2, \ldots, m \qquad (3.3)$$

where all x_{ij}'s are in a common scale.

One way of making a common scale is using the degree of closeness to the ideal solution [459]; the ratio of an attribute value to the most preferable attribute value ($x_j^* = \max_i x_{ij}$), that is

$$r_{ij} = \frac{x_{ij}}{x_j^*} \qquad (3.4)$$

provided that attribute j is a benefit criterion (i.e., the larger x_j, the more preference). A more complicated form of r_{ij} is (see eq. (2.5))

$$r_{ij} = \frac{x_{ij} - x_j^{min}}{x_j^* - x_j^{min}} \qquad (3.5)$$

where $x_j^* = \max_i x_{ij}$, $\qquad i = 1,2,\ldots,m$

$x_j^{min} = \min_i x_{ij}$, $\qquad i = 1,2,\ldots,m$

Then the maximin procedure becomes

$$\max_i \min_j r_{ij}, \qquad j = 1,2,\ldots,n; \qquad i = 1,2,\ldots,m \qquad (3.6)$$

Note that in case of a cost criterion, r_{ij} has to be computed as (see eq. (2.4) or eq. (2.6))

$$r_{ij} = \frac{1/x_{ij}}{\max_i (1/x_{ij})} = \frac{\min x_{ij}}{x_{ij}} = \frac{x_j^{min}}{x_{ij}} \qquad (3.7)$$

or

$$r_{ij} = \frac{x_j^* - x_{ij}}{x_j^* - x_j^{min}} \qquad (3.8)$$

Numerical Example (The Fighter Aircraft Problem)

The decision matrix after assigning numerical values (by using the interval scale of Fig. 2.4) to qualitative attributes is

$$D = \begin{array}{c} \\ \\ \\ \\ \end{array} \begin{array}{cccccc} X_1 & X_2 & X_3 & X_4 & X_5 & X_6 \\ \left[\begin{array}{cccccc} 2.0 & 1500 & 20,000 & 5.5 & 5 & 9 \\ 2.5 & 2700 & 18,000 & 6.5 & 3 & 5 \\ 1.8 & 2000 & 21,000 & 4.5 & 7 & 7 \\ 2.2 & 1800 & 20,000 & 5.0 & 5 & 5 \end{array}\right] & \begin{array}{c} A_1 \\ A_2 \\ A_3 \\ A_4 \end{array} \end{array}$$

The maximin procedure requires a common scale for all elements in the decision matrix. The converted decision matrix, by using eq. (3.7) for attribute X_4 and eq. (3.4) for all other attributes, is

$$D' = \begin{bmatrix} 0.80 & 0.56 & 0.95 & 0.82 & 0.71 & 1.0 \\ 1.0 & 1.0 & 0.86 & 0.69 & 0.43 & 0.56 \\ 0.72 & 0.74 & 1.0 & 1.0 & 1.0 & 0.78 \\ 0.88 & 0.67 & 0.95 & 0.90 & 0.71 & 0.56 \end{bmatrix} \begin{matrix} A_1 \\ A_2 \\ A_3 \\ A_4 \end{matrix}$$

with column headers X_1, X_2, X_3, X_4, X_5, X_6.

In applying the maximin procedure to this example, we note that the lowest attribute value for A_1 is 0.56 (for X_2, the ferry range); the lowest attribute value for A_2 is 0.43 (for X_5, the reliability); the lowest attribute value for A_3 is 0.72 (for X_1, the maximum speed); and the lowest attribute value for A_4 is 0.36 (for X_6, the maneuverability). Since the largest of these four minimum values is 0.72, alternative A_3 would be selected because it has the maximum-minimum value.

Note

This method utilizes only a small part of the available information in making a final choice -- only one attribute per alternative. Thus even if an alternative is clearly superior in all but one attribute which is below average, another alternative with only average on all attributes would be chosen over it. The maximin method, then, has some obvious short-comings in decision making. What is appropriate for selecting a chain is not necessarily appropriate for general decision making [273]. The procedure is reasonable for selecting a chain because all the links are used at the same time and they are essentially interchangeable, in that no link is worth more, or performs a different function, than some other link.

The applicability of the maximin method is relatively limited. In general the method would be reasonable only if the DM assumed to have pessimistic nature in the decision making situation. The maximin and its reverse, the minimax procedure, is widely used in game theory [421].

1.1.3 MAXIMAX

In contrast to the maximin method, the maximax method selects an alter-
nate by its best attribute value rather than its worst attribute value [273].
In this case the highest attribute value for each alternative is identified,
then these maximum values are compared in order to select the alternative
with the largest such value, the maximax procedure.

Note that in this procedure, as with the maximin procedure only the single
strongest attribute represents an alternative; all other (n-1) attributes
for the particular alternative are ignored; and it may evaluate different
attributes in a final choice among alternatives.   Therefore, as with the
maximin method, the maximax method can be used only when all attributes are
measured on a common scale (see eqs. (3.3), (3.4), (3.7) and (3.8)).   The
alternative , A^+, is selected such that

$$A^+=\{A_i \mid = \max_i \max_j r_{ij}\}, \qquad j = 1,2,\ldots,n; \qquad i = 1,2,\ldots,m \qquad (3.9)$$

The comparability assumptions and incompleteness properties of the maximax method do not make it a very useful technique for general decision
making [273]. However, just as the maximin method may have a domain in which
it is quite reasonable, the maximax method may also be reasonable in some
specific decision making situations.   As an example, pro-football teams use
the maximax procedure to draft players.

Numerical Example (The Fighter Aircraft Problem)

The decision matrix converted to a common scale is

$$D' = \begin{array}{cccccc} X_1 & X_2 & X_3 & X_4 & X_5 & X_6 \\ \left[\begin{matrix} 0.80 & 0.56 & 0.95 & 0.82 & 0.71 & 1.0 \\ 1.0 & 1.0 & 0.86 & 0.69 & 0.43 & 0.56 \\ 0.72 & 0.74 & 1.0 & 1.0 & 1.0 & 0.78 \\ 0.88 & 0.67 & 0.95 & 0.90 & 0.71 & 0.36 \end{matrix}\right] & \begin{matrix} A_1 \\ A_2 \\ A_3 \\ A_4 \end{matrix} \end{array}$$

In applying the maximax procedure to this example, the highest attribute value for A_1 is 1.0 (for X_6, the maneuverability); the highest attribute value for A_2 is 1.0 (for both X_1, the maximum speed, and X_2, the ferry range); the highest attribute value for A_3 is 1.0 (for X_3, X_4, and X_5, the maximum payload, acquisition cost, and reliability, respectively); and the highest attribute value for A_4 is 0.95 (for X_3, the maximum payload). Maximizing these maximum values would lead then to a choice of A_1, A_2, or A_3.

Note

Both the maximin procedure and the maximax procedure use what could be called a specialized degenerate weighting, which may be different for each alternative [299]: the maximin method assigns a weight of 1 to the worst attribute value and a weight of 0 to all others; the maximax method assigns a weight of 1 to the best attribute value and a weight of 0 to all others.

The maximin procedure describes the individual as finding the worst outcome (attribute value) for each choice (alternative), and then choosing that choice (alternative) for which the worst outcome is least bad. This policy is ultra-pessimistic in that, for each possible choice, it considers only the worst that can happen, while ignoring other aspects of the choice. The maximax procedure, in contrast, is ultra-optimistic, in that it describes the individual as choosing that choice for which the best outcome is the best. The Hurwicz procedure [178a] is an amalgamation of the above two, in that it takes into account both the worst and the best. It selects A^+ such that

$$(3.10)$$

$$A^+ = \{A_i \mid \max_i \, [\, \alpha \, \min_j \, r_{ij} + (1 - \alpha) \, \max_j \, r_{ij} \,]\}$$

The weight α is referred to as the pessimism-optimism index; it is supposed to vary (over $0 \leq \alpha \leq 1$) among the individual DMs; the higher α the more pessimistic the individual DM. As is apparent, the extreme case $\alpha = 1$ gives the maximin, while $\alpha = 0$ the maximax.

Although this procedure might seem useful in a single instance, it is clearly inadequate when considering the whole multiple attribute problem (e.g., drafting the whole team on the basis of a single attribute -- unless you want a team of kickers) [275].

2. METHODS FOR INFORMATION ON ATTRIBUTE GIVEN

A DM may express his/her preference information either on attributes or on alternatives. Usually the information on attributes is less demanding to assess than that on alternatives. The majority of MADM methods require this kind of information to process inter- and intra-attribute comparisons.

The information can be expressed in various ways: 1) standard level of each attribute, 2) relative importance of each attribute by ordinal preference, 3) relative importance of each attribute by cardinal preference, and 4) marginal rate of substitution (MRS) between attributes. Standard levels or ordinal preference information is utilized in the noncompensatory models, and cardinal preference or marginal rate of substitution is needed in the compensatory models.

2.1 METHODS FOR STANDARD LEVEL OF ATTRIBUTE GIVEN

To obtain a driving license one must meet certain absolute standards. The DM sets up the minimal attribute values (standard levels) he/she will accept for each of the attributes. Any alternative (or candidate) which has an attribute value less than the standard level will be rejected. This procedure is called the conjunctive method [70] or the satisficing method described by Simon [373]. On the other hand, if evaluation of an alternative (or candidate) is based upon the greatest value of only one attribute, the procedure is called the disjunctive method. Selection of professional football players is done according to this method. (See Table 1.3 for references.)

2.1.1 CONJUNCTIVE METHOD (SATISFICING METHOD)

Consider, for example, the position of a visiting American history teacher in a French school [70]. An individual's effectiveness as a teacher will be limited by the lesser of his/her abilities in history and French; he/she cannot compensate for an insufficient knowledge of French by an excellent knowledge of history, or vice versa. The school wants to eliminate the candidates who do not possess the acceptable knowledge in both fields. In the conjunctive method (or satisficing method), all the standards must be passed in order for the alternatives to be acceptable.

To apply the method, the DM must supply the minimal attribute values (the cutoff values) acceptable for each of the attributes. The cutoff values given by the DM play the key role in eliminating the noncontender alternatives; if too high, none is left; if relatively low quite a few alternatives are left after filtering. Hence increasing the minimal standard levels in an iterative way, we can sometimes narrow down the alternatives to a single choice.

We classify A_i as an acceptable alternative only if

$$x_{ij} \geq x_j^0, \qquad j = 1, 2, \ldots, n \qquad\qquad (3.11)$$

where x_j^0 is the standard level of x_j.

Numerical Example (The Fighter Aircraft Problem)

In the fighter aircraft problem, suppose that the DM specified the following minimal requirements (the cutoff values): $\underline{x}^0 = (2.0, 1500, 20,000,$ 6.0, average, average). Given these minimal acceptable values, alternatives A_1 and A_4 are acceptable; that is, they satisfy these requirements. Alternative A_2 fails to satisfy them because its maximum payload (X_3) is too low (18,000 vs. 20,000), its acquisition cost (X_4) is too high (6.5 vs. 6.0), and its reliability (X_5) is too low (low vs. average); alternative A_3 fails because its maximum speed (X_1) is too low (1.8 vs. 2.0).

Note

The conjunctive (satisficing) method is not usually used for selection of alternatives but rather for dichotomizing them into acceptable/not acceptable categories. Dawes [70] developed a way to set up the standards if the DM wants to dichotomize the alternatives.

Consider a set of n equally weighted independent attributes. Let

    r = the proportion of alternatives which are rejected,

    P_c = the probability that a randomly chosen alternative scores above the conjunctive cutting level.

Then

$$r = 1 - P_c^n \qquad\qquad (3.12)$$

since the probability of being rejected is equal to one minus the probability of passing on all attributes. From eq. (3.12), we obtain

$$P_c = (1 - r)^{1/n} \tag{3.13}$$

For example [70], suppose that a college evaluates applicants for admissions on each of four attributes - intellectual ability, academic ability, extracurricular activities, and character - and that it has applicants' scores on each of these attributes. Assume that these attributes are independent and that the college considers them all equally important. Suppose that the college wishes to accept one fifth of its applicants. Now we have

$$n = 4$$

$$r = \frac{4}{5}$$

$$P_c = (1 - \frac{4}{5})^{\frac{1}{4}} = 0.67$$

Hence, the college must choose a cutting score for each attribute such that 67% of its applicants will place above this score. This cutting score is called the conjunctive cutting score.

The conjunctive method does not require that the attribute information be in numerical form, and information on the relative importance of the attributes is not needed. The method is noncompensatory; if we simply use minimum cutoff values for each of the attributes, none of the alternative systems gets credited for especially good attribute values. Thus A_3 of the fighter aircraft problem fails even though it has the lowest acquisition cost. The attempts to credit alternatives with especially high values suggest other methods to be discussed later.

Because of its strong intuitive appeal, the conjunctive method (or satisficing method) has long been used.

2.1.2 DISJUNCTIVE METHOD

A disjunctive method is one in which an alternative (or an individual) is evaluated on its greatest value (or talent) of an attribute. For example [70], professional football players are selected according to the disjunctive method; a player is selected because he can either pass exceptionally, or run exceptionally, or kick exceptionally, etc. A player's passing ability is irrelevant if he is chosen for his kicking ability.

We classify A_i as an acceptable alternative only if

$$x_{ij} \geq x_j^0, \qquad j = 1 \text{ or } 2 \text{ or } \dots \text{ or } n \tag{3.13}$$

where x_j^0 is a desirable level of x_j.

A disjunctive method guarantees selection of all individuals (candidates) with any extreme talent, while the conjunctive method guarantees rejection of all individuals with an extremely small talent.

Numerical Example (The Fighter Aircraft Problem)

In the fighter aircraft problem, suppose that the DM specified the following desirable levels for each attribute: $\underline{x}^0$ = (2.4, 2500, 21,000, 4.5, very high, very high). Given these desirable levels, alternatives A_1, A_2, and A_3 are acceptable, and alternative A_4 is rejected. A_1 is accepted because its maneuverability (X_6) is acceptable (very high vs. very high); A_2 is accepted because its maximum speed (X_1) is acceptable (2.5 vs. 2.4), and/or its ferry range (X_2) is acceptable (2700 vs. 2500); and A_3 is accepted because its maximum payload (X_3) is acceptable (21,000 vs. 21,000), and/or its acquisition cost (X_4) is acceptable (4.5 vs. 4.5). A_4 is rejected because it does not possess any one of the desirable levels specified by the DM.

Note

For the disjunctive method, the probability of being rejected is equal to the probability of failing on all attributes:

$$r = (1 - P_d)^n \qquad\qquad (3.14)$$

where r is the proportion of alternatives which are rejected, and P_d is the probability that a randomly chosen alternative scores above the disjunctive cutting level. From eq. (3.14), we obtain

$$P_d = 1 - r^{1/n} \qquad\qquad (3.15)$$

For example, in the problem of college evaluation of applicant admission which we considered in the section on the conjunctive method (2.1.1), if the college uses a disjunctive method, it will accept any applicant who scores above the cutting score on any attribute.

Then we have

$$n = 4$$

$$r = \frac{4}{5}$$

$$P_d = 1 - (\frac{4}{5})^{1/4} = 0.05$$

that is, the disjunctive probability, P_d, equals 0.05, hence the disjunctive cutting score for each attribute will be one such that 5% of the applicants score above it -- a contrast to the conjunctive cutting score of 67% [70].

As with the conjunctive method, the disjunctive method does not require that the attribute information be in numerical form, and it does not need information on the relative importance of the attributes.

2.2. METHODS FOR ORDINAL PREFERENCE OF ATTRIBUTE GIVEN

The most important information needed for lexicographic method, elimination by aspects, and permutation method is the ordinal interattribute preference information. The relative importance among attributes determined by ordinal preference is less demanding for the DM to assess than that by cardinal preference.

The permutation method was originally developed for the cardinal preferences of attributes given, but it is better used for the ordinal preferences given, and the method will identify the best ordering of the alternative rankings.

2.2.1   LEXICOGRAPHIC METHOD

In some decision situations a single attribute seems to predominate. For example, "buy the cheapest" rule is that in which the price is the most important attribute to that DM. One way of treating this situation is to compare the alternatives on the most important attribute. If one alternative has a higher attribute value than any of the other alternatives, the alternative is chosen and the decision process ends. However, if some alternatives are tied on the most important attribute, the subset of tied alternatives are then compared on the next most important attribute. The process continues sequentially until a single alternative is chosen or until all n attributes have been considered.

The method requires that the attributes be ranked in the order of importance by the DM. Let the subscripts of the attributes indicate not only the components of the attribute vector, but also the priorities of the attributes, i.e., X_1 be the most important attribute to the DM, X_2 the second most important one, and so on. Then alternative(s), A^1, is(are) selected such that

$$A^1 = \{A_i \mid \max_i x_{i1}\}, \qquad i = 1, 2, \ldots, m \tag{3.16}$$

If this set $\{A^1\}$ has a single element, then this element is the most preferred alternative. If there are multiple maximal alternatives, consider

$$A^2 = \{A^1 \mid \max_i x_{i2}\}, \qquad i \in \{A^1\} \tag{3.17}$$

If this set $\{A^2\}$ has a single element, then stop and select this alternative. If not, consider

$$A^3 = \{A^2 \mid \max_i x_{i3}\}, \qquad i \in \{A^2\} \tag{3.18}$$

Continue this process until either (a) some $\{A^k\}$ with a single element is found which is then the most preferred alternative, or (b) all n attributes have been considered, in which case, if the remaining set contains more than one element, they are considered to be equivalent.

LEXICOGRAPHIC SEMIORDER

The lexicographic semiorder, described by Luce [269] and Tversky [409] is closely related to the lexicography. In most cases it makes sense to allow bands of imperfect discrimination so that one alternative is not judged better just because it has a slightly higher value on one attribute. In a lexicographic semiorder, a second attribute is considered not only in cases where values for several alternatives on the most important attribute are equal but also for cases where the differences between the values on the most important attribute are not significant or noticeable. This same process may then be used for further attributes if more than one alternative still remain. Thus a consideration of whether differences are significant is imposed upon lexicographic ordering.

Tversky [409] showed how this method can lend itself to intransitive choices in which A_1 is chosen over A_2, A_2 over A_3, and A_3 over A_1. Consider, for example, the following decision matrix with two benefit attributes,

$$D = \begin{matrix} & X_1 & X_2 & \\ & \begin{bmatrix} 2 & 6 \\ 3 & 4 \\ 4 & 2 \end{bmatrix} & \begin{matrix} A_1 \\ A_2 \\ A_3 \end{matrix} \end{matrix}$$

Suppose that X_1 is the more important attribute, but that differences of one or less are assumed to be not significant. Also suppose that differences in X_2 of one or less are viewed to be not significant. The A_1 will be preferred to A_2 (since difference of X_1 is not significant but difference of X_2 is), A_2 to A_3, but A_3 to A_1 (since difference of X_1 is significant).

Numerical Example (The Fighter Aircraft Problem)

In the fighter aircraft problem, suppose that X_1 is the most important attribute, and then X_3, X_2,..., in that order. Then A_2 is chosen by the lexicographic method because its maximum speed (X_1) of 2.5 Mach is the largest among the alternatives.

It is also assumed that 0.3 Mach or less difference in X_1 is not significant, and that 1000 pounds or less difference in X_3 (the maximum payload) is not significant. Then by the lexicographic semiorder A_4 is selected as the solution because the difference of X_1 between A_2 and A_4 is not significant ($x_{21} = 2.5$ vs $x_{41} = 2.2$), but the difference of X_3 between them is significant ($x_{23} = 18000$ vs $x_{43} = 20000$).

Note

The lexicographic method is also used in multiple objective decision making (MODM) (see Hwang and Masud [BM-11]). Theoretical considerations of the method are given by Luce [269], Shepard [367], and Fishburn [120].

This method as it is in maximin (1.1.2) and maximax (1.1.3) utilizes only a small part of the available information in making a final choice. For example, for the fighter aircraft problem only the information of X_1 and X_3 are used. Lexicography is somewhat more demanding of information than maximin and maximax, because it requires a ranking of the importance of the attributes, whereas maximin and maximax do not. However, lexicography does not require comparability across attributes as did maximin and maximax.

When applied to general decision making, the lexicographic method requires information on the preference among attribute values and the order in which attributes should be considered. In both cases, it needs only ordering or ranking information and not (necessarily) numerical values. Because of its limited information requirements, lexicography has received serious consideration as a decision technique in a number of areas (see MacCrimmon [273]).

2.2.2 ELIMINATION BY ASPECTS

Elimination by aspects is a classical human decision making procedure which has been used for a long time. Tversky [410, 410a, 410b] formalizes the decision process mathematically with the introduction of choice probability as a theoretical concept in the analysis of choice.

The DM, as in conjunctive method (2.1.1), is assumed to have minimum cutoffs for each attribute. An attribute is selected, and all alternatives not passing the cutoff on that attribute are eliminated. Then another attribute is selected, and so forth. The process continues until all alternatives but one are eliminated [29]. Like lexicographic method (2.2.1), it examines one attribute at a time, making comparisons among alternatives. However, it does differ slightly since it eliminates alternatives which do not satisfy some standard level, and it continues until all alternatives except one have been eliminated. Another difference is that the attributes are not ordered in terms of importance, but in terms of their discrimination power in a probabilistic mode.

In Tversky's model, each alternative is viewed as a set of aspects. The aspects could represent values along some fixed quantitative or qualitative dimensions (attributes) (e.g., price, quality, comfort), or they could be arbitrary features of the alternatives that do not fit into any simple dimensional structure. Since the model describes choice as an elimination process governed by successive selection of aspects instead of cutoffs, it is called the Elimination by Aspects (EBA).

As an illustration of elimination by aspects, Tversky [410] describes a television commercial   that advertises a computer course:

"There are more than two dozen companies in the San

Francisco area which offer training in computer programming."

The announcer puts some two dozen eggs and one walnut on the table to represent the alternatives, and continues:

"Let us examine the facts. How many of these schools

have on-line computer facilities for training?"

The announcer removes several eggs.

"How many of these schools have placement services

that would help find you a job?"

The announcer removes some more eggs.

"How many of these schools are approved for veterans'

benefits?"

This continues until the walnut alone remains. The announcer

cracks the nutshell, which reveals the name of the company and

concludes:

"This is all you need to know in a nutshell."

In the above example, alternatives are eliminated by the aspects of "on-

line computer facilities for training," "placement services," "approval for

veterans' benefits," etc.

The rationale of EBA is briefly described here. We begin by intro-

ducing some notation. Let $T = \{x,y,z,...\}$ be a finite set, interpreted

as the total set of alternatives under consideration. We use A, B, C, ...,

to denote specific nonempty subsets of T. The probability of choosing

an alternative x from an offered set $A \subseteq T$ is denoted $P(x,A)$. Naturally,

we assume $P(x,A) \geq 0$, $\sum_{x \in A} P(x,A) = 1$ for any A, and $P(x,A) = 0$ for any

$x \notin A$. For brevity, we write $P(x;y)$ for $P(x,\{x,y\})$, $P(x; y,z)$ for $P(x,\{x,y,z\})$,

etc. To formalize the method we introduce the aspects in an explicit fashion.

Let M be a mapping that associates with each $x \in T$ a nonempty set $M(x) = x' =$

$\{\alpha, \beta, \gamma, ...\}$ of elements which are interpreted as the aspects of X.

To clarify the formalization of the method let us first look at a simple example [410a]. Consider a three-alternative set T = {x,y,z}, where the collections of aspects associated with the respective alternatives are

$$x' = \{\alpha_1, \ \alpha_2, \theta_1, \theta_2, \rho_1, \rho_2, \omega\}$$
$$y' = \{\beta_1, \beta_2, \theta_1, \theta_2, \sigma_1, \sigma_2, \omega\}$$

and

$$z' = \{\gamma_1, \gamma_2, \rho_1, \rho_2, \sigma_1, \sigma_2, \omega\}$$

A graphical representation of the structure of the alternatives and their aspects is presented in Fig. 3.2. It is readily seen that α_i, β_i, and $\gamma_i (i = 1, 2)$ are, respectively, the unique aspects of x, y, and z; that θ_i, σ_i, and ρ_i are, respectively, the aspects shared by x and y, by y and z, and by x and z; and that ω is shared by all three alternatives. Since the selection of ω does not eliminate any alternative, it can be discarded from further consideration. Let u be a scale which assigns to each aspect a positive number representing its utility or value, and let K be the sum of the scale values of all the aspects under consideration, that is, $K = \Sigma_\alpha u(\alpha)$ where the summation ranges over all the aspects except ω. Using these notations we now compute P(x,T). Note first that x can be chosen directly from T if either α_1 or α_2 is selected in the first stage (in which case both y and z are eliminated). This occurs with probability $[u(\alpha_1) + u(\alpha_2)]/K$. Alternatively, x can be chosen via {x,y} if either θ_1 or θ_2 is selected in the first stage (in which case z is eliminated), and then x is chosen over y. This occurs with probability $[u(\theta_1) + u(\theta_2)] \cdot P(x; y)/K$. Finally, x can be chosen via {x,z} if either ρ_1 or ρ_2 is selected in the first stage (in which case y is eliminated), and then x is chosen over z. This occurs with probability

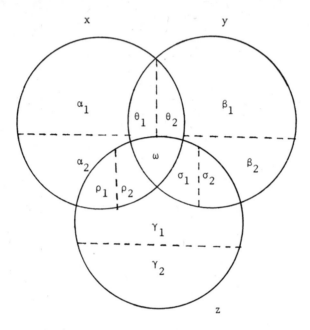

Fig. 3.2 A graphical representation of aspects in the three-
alternative case [410a].

$[u(\rho_1) + u(\rho_2)]P(x;z)/K$. Since the above paths leading to the choice

of x from T are all disjoint,

$$P(x,T) = (1/K)(u(\alpha_1) + u(\alpha_2)$$

$$+ [u(\theta_1) + u(\theta_2)]P(x;y)$$

$$+ [u(\rho_1) + u(\rho_2)]P(x;z))$$

where

$$P(x;y) = \frac{u(\alpha_1) + u(\alpha_2) + u(\rho_1) + u(\rho_2)}{u(\alpha_1) + u(\alpha_2) + u(\rho_1) + u(\rho_2) + u(\beta_1) + u(\beta_2) + u(\sigma_1) + u(\sigma_2)} , \text{ etc.}$$

The three-alternative case can be extended. In general, let T be

any finite set of alternatives. For any $A \subseteq T$ let $A' = \{\alpha | \alpha \epsilon x' \text{ for some }$

$x \epsilon A\}$, and $A^o = \{\alpha | \alpha \epsilon x' \text{ for all } x \epsilon A\}$. Thus, A' is the set of aspects

that belongs to at least one alternative in A, and A^o is the set of aspects

that belongs to all the alternatives in A. In particular, T' is the set

of all aspects under consideration, while T^o is the set of aspects shared

by all the alternatives under study. Given any aspect $\alpha \epsilon T'$, let A_α denote

those alternatives of A which include α, that is, $A_\alpha = \{x | x \epsilon A \text{ and } \alpha \epsilon x'\}$.

The EBA asserts that there exists a positive scale u defined on the aspects

(or more specifically on $T' - T^o$) such that for all $x \epsilon A \subseteq T$

$$P(x,A) = \frac{\sum_{\alpha \epsilon x' - A^o} u(\alpha)P(x,A_\alpha)}{\sum_{\beta \epsilon A' - A^o} u(\beta)} \qquad (3.19)$$

provided the denominator does not vanish. Note that the summations in the

numerator and the denominator of eq. (3.19) range, respectively, over all

aspects of x and A except those that are shared by all elements of A. Hence, the denominator of eq. (3.19) vanishes only if all elements of A share the same aspects, in which case it is assumed that $P(x,A) = 1/a$, where a is the number of elements in A.

Eq. (3.19) is a recursive formula. It expresses the probability of choosing x from A as a weighted sum of the probabilities of choosing x from the various subsets of A (i.e., A_α for $\alpha \in x'$), where the weights (i.e., $u(\alpha)/\Sigma \ u(\beta)$) correspond to the probabilities of selecting the re-spective aspects of x.

Tversky expands eq. (3.19) (where the scale u is defined over individual aspects) into the choice probability where a scale can be defined over collections of aspects which are associated, respectively, with the subsets of T. This expansion does not require any prior characterization of the alternatives in terms of their aspects.

Note

The EBA is similar in some respects to the lexicographic method (2.2.1) and conjunctive method (2.1.1). It differs from them in that, due to its probabilistic nature, the criteria for elimination (i.e., the selected aspects) and the order in which they are applied vary from one occasion to another and are not determined in advance. And the lexicographic method does not require a set of aspects (i.e., standard level or desirable quality). It also differs from conjunctive method in that the number of criteria for elimination varies. If an aspect which belongs only to a single alternative is chosen at the first stage, then EBA needs only one aspect.

The EBA has some advantages: it is relatively easy to apply; it involves no numerical computations, and it is easy to explain and justify in terms of a priority ordering defined on the aspects. The major flow on the logic of EBA lies in the noncompensatory nature of the selection process. Although each selected aspect is desirable, it might lead to the elimination of alternatives that are better than those which are retained. In general, the strategy of EBA cannot be defended as a rational procedure of choice. On the other hand, there may be many contexts in which it provides a good approximation to much more complicated compensatory models and could thus serve as a useful simplication procedure.

2.2.3 PERMUTATION METHOD

The permutation method [315] uses Jaquet-Lagrèze's successive permutations of all possible rankings of alternatives. The method consists of testing each possible ranking of the alternatives against all others. With m alternatives, m! permutation rankings are available. The method will identify the best ordering of the alternative rankings, then the dominating alternative. The method was originally developed for the cardinal preferences of attributes (i.e., a set of weight) given, but it is rather to be used for the ordinal preferences given. We will discuss first the case of cardinal preferences given, then the ordinal ones.

Suppose a number of alternatives $(A_i, i = 1,2,...,m)$ have to be evaluated according to n attributes $(X_j, j = 1, 2,...,n)$. The problem can be stated in a decision matrix D:

$$
D = \begin{array}{c} \\ A_1 \\ A_2 \\ \vdots \\ A_m \end{array}
\begin{array}{cccc}
X_1 & X_2 & \cdots & X_n \\
\left[\begin{array}{cccc}
x_{11} & x_{12} & \cdots & x_{1n} \\
x_{21} & x_{22} & \cdots & x_{2n} \\
\vdots & \vdots & & \\
x_{m1} & x_{m2} & \cdots & x_{mn}
\end{array}\right]
\end{array}
$$

Assume that a set of cardinal weights w_j, $j = 1,2,...,n$, $\sum_j w_j = 1$, be given to the set of corresponding attributes.

Suppose that there are three alternatives: A_1, A_2, and A_3. Then six permutations of the ranking of the alternatives exist (m! = 3! = 6). They are:

$$P_1 = (A_1, A_2, A_3) \qquad\qquad P_4 = (A_2, A_3, A_1)$$

$$P_2 = (A_1, A_3, A_2) \qquad\qquad P_5 = (A_3, A_1, A_2)$$

$$P_3 = (A_2, A_1, A_3) \qquad\qquad P_6 = (A_3, A_2, A_1)$$

Assume that a testing order of the alternatives be $P_5 = (A_3, A_1, A_2)$. Then the set of concordance partial order is $\{A_3 \geq A_1, A_3 \geq A_2, A_1 \geq A_2\}$, and the set of discordance is $\{A_3 \leq A_1, A_3 \leq A_2, A_1 \leq A_2\}$.

If in the ranking the partial ranking $A_k \geq A_\ell$ appears, the fact that $x_{kj} \geq x_{\ell j}$ will be rated w_j, $x_{kh} \leq x_{\ell h}$ being rated $-w_h$. The evaluation criterion of the chosen hypothesis (ranking of the alternatives) is the algebraic sum of w_j's corresponding to the element by element consistency. Consider the ith permutation:

$$P_i = (\ldots, A_k, \ldots, A_\ell, \ldots), \qquad i = 1, 2, \ldots, m!$$

where A_k is ranked higher than A_ℓ. Then the evaluation criterion of P_i, R_i, is given by

$$R_i = \sum_{j \in C_{k\ell}} w_j - \sum_{j \in D_{k\ell}} w_j, \qquad i = 1, 2, \ldots, m! \tag{3.20}$$

where

$$C_{k\ell} = \{j \,|\, x_{kj} \geq x_{\ell j}\}, \qquad k, \ell = 1, 2, \ldots, m, \; k \neq \ell$$

$$D_{k\ell} = \{j \,|\, x_{kj} \leq x_{\ell j}\}, \qquad k, \ell = 1, 2, \ldots, m, \; k \neq \ell$$

The concordance set $C_{k\ell}$ is the subset of all criteria for which $x_{kj} \geq x_{\ell j}$, and the discordance set $D_{k\ell}$ is the subset of all criteria for which $x_{kj} \leq x_{\ell j}$.

Numerical Example (The Fighter Aircraft Problem)

The decision matrix of the fighter aircraft problem is:

$$D = \begin{array}{c c c c c c c c}
 & X_1 & X_2 & X_3 & X_4 & X_5 & X_6 & \\
 \begin{bmatrix} 2.0 & 1500 & 20000 & 5.5 & \text{average} & \text{very high} \end{bmatrix} & A_1 \\
 2.5 & 2100 & 18000 & 6.5 & \text{low} & \text{average} & A_2 \\
 1.8 & 2000 & 21000 & 4.5 & \text{high} & \text{high} & A_3 \\
 2.2 & 1800 & 20000 & 5.0 & \text{average} & \text{average} & A_4
\end{array}$$

Assume that the cardinal weight of the attributes be $\underline{w}$ = (0.2, 0.1, 0.1, 0.1, 0.2, 0.3). There are 24 (=4!) permutations of the alternatives which have to be tested. They are:

$P_1 = (A_1, A_2, A_3, A_4)$ $P_{13} = (A_3, A_1, A_2, A_4)$

$P_2 = (A_1, A_2, A_4, A_3)$ $P_{14} = (A_3, A_1, A_4, A_2)$

$P_3 = (A_1, A_3, A_2, A_4)$ $P_{15} = (A_3, A_2, A_1, A_4)$

$P_4 = (A_1, A_3, A_4, A_2)$ $P_{16} = (A_3, A_2, A_4, A_1)$

$P_5 = (A_1, A_4, A_2, A_3)$ $P_{17} = (A_3, A_4, A_1, A_2)$

$P_6 = (A_1, A_4, A_3, A_2)$ $P_{18} = (A_3, A_4, A_2, A_1)$

$P_7 = (A_2, A_1, A_3, A_4)$ $P_{19} = (A_4, A_1, A_2, A_3)$

$P_8 = (A_2, A_1, A_4, A_3)$ $P_{20} = (A_4, A_1, A_3, A_2)$

$P_9 = (A_2, A_3, A_1, A_4)$ $P_{21} = (A_4, A_2, A_1, A_3)$

$P_{10} = (A_2, A_3, A_4, A_1)$ $P_{22} = (A_4, A_2, A_3, A_1)$

$P_{11} = (A_2, A_4, A_1, A_3)$ $P_{23} = (A_4, A_3, A_1, A_2)$

$P_{12} = (A_2, A_4, A_3, A_1)$ $P_{24} = (A_4, A_3, A_2, A_1)$

Let us, for example, compute the testing of the ordering $P_4 = (A_1, A_3, A_4, A_2)$ derived from the decision matrix, D. The testing result of P_4 is presented in matrix C_4.

$$
C_4 = \begin{array}{c} \\ 1 \\ 3 \\ 4 \\ 2 \end{array}
\begin{array}{cccc}
1 & 3 & 4 & 2 \\
\left[\begin{array}{cccc}
0 & 0.5 & 0.6 & 0.7 \\
0.5 & 0 & 0.8 & 0.7 \\
0.7 & 0.2 & 0 & 0.7 \\
0.3 & 0.6 & 0.3 & 0
\end{array}\right]
\end{array}
$$

The figures in the cells have been computed as follows:

$$c_{13} = \sum_{j \varepsilon C_{13}} w_j = w_1 + w_6 = 0.2 + 0.3 = 0.5$$

where $x_{11}(= 2.0) > x_{31}(= 1.8)$ that gives w_1, and $x_{16}(= $ very high$) > x_{36}$ ($=$ high) that gives w_6.

$$c_{31} = \sum_{j \varepsilon D_{13}} w_j = w_2 + w_3 + w_4 + w_5 = 0.1 + 0.1 + 0.1 + 0.2 = 0.5$$

where $x_{12}(= 1500) < x_{32}(= 2000)$ that gives $-w_2$, $x_{13}(= 20,000) < x_{33}$ ($= 21,000$) that gives $-w_3$, $x_{14}(= 5.5$ cost$) < x_{34}(= 4.5$ cost$)$ that gives $-w_4$, and x_{15} ($=$ average$) < x_{35}(= $ high$)$ that gives $-w_5$. Then the evaluating criterion of P_4, R_4, is:

$$R_4 = \sum_{j \varepsilon C_{k\ell}} w_j - \sum_{j \varepsilon D_{k\ell}} w_j = 4.0 - 2.6 = 1.4$$

where $\sum_{j \varepsilon C_{k\ell}} w_j$ is the sum of the upper-triangular elements of matrix C_4 in accordance (concordance) with the hypothesis: $A_1 > A_3 > A_4 > A_2$, and

$\sum_{j \varepsilon D_{k\ell}} w_j$ is the

sum of the lower-triangular elements in conflict (discordance) with the hypothesis.

Similarly, 24 different C_i, $i = 1,2,\ldots,24$ matrices are computed and their evaluation criteria, R_i, $i = 1,2,\ldots,24$ are calculated. They are:

$$\underline{R} = \{ R_1 = 0.4, \ R_2 = -0.6, \ R_3 = 1.2, \ R_4 = 1.4, \ R_5 = -0.6, \ R_6 = 0.2,$$

$$R_7 = -0.4, \ R_8 = -1.6, \ R_9 = -0.4, \ R_{10} = -0.2, \ R_{11} = -1.4, \ R_{12} = -1.4,$$

$$R_{13} = 1.2, \ R_{14} = 1.4, \ R_{15} = 0.4, \ R_{16} = 0.6, \ R_{17} = 1.6, \ R_{18} = 0.8,$$

$$R_{19} = -0.4, \ R_{20} = 0.4, \ R_{21} = -1.2, \ R_{22} = -1.2, \ R_{23} = 0.4, \ R_{24} = -0.4 \}$$

Since $R_{17} = 1.6$ gives the maximum value, the best order of the alternatives is $P_{17} = (A_3, A_4, A_1, A_2)$.

Case of a Qualitative Attribute Evaluation (with Ordinal Importance of Attribute Given): It is easy to notice that the four sets of weights:
$$\underline{w}^1 = (1, 0, 0, 0), \ \underline{w}^2 = (\tfrac{1}{2}, \tfrac{1}{2}, 0, 0), \ \underline{w}^3 = (\tfrac{1}{3}, \tfrac{1}{3}, \tfrac{1}{3}, 0), \ \underline{w}^4 = (\tfrac{1}{4}, \tfrac{1}{4}, \tfrac{1}{4}, \tfrac{1}{4}),$$
satisfy the ordered weight of $(w_1 \geq w_2 \geq w_3 \geq w_4)$. Through this relationship Paelinck [315] gives the following theorem on which the algorithm for cases of a qualitative attribute (or with an ordinal weight) given.

THEOREM: The admissible field of weights defined by

$$\sum_j w_j = 1, \qquad j = 1,2,\ldots,n$$

$$w_k \geq w_\ell, \qquad k < \ell$$

can be subdivided into subfields on the sole basis of the evaluation of each ranking for the extreme points $(1, 0)$, $(0.5, 0.5, 0)$, $(0.333, 0.333, 0.333, 0)$, ...

The algorithm based on this theorem is quite simple. Evaluate each ranking for the successive end points $(1, 0)$, $(0.5, 0.5, 0)$, $\ldots$, and select the maximum values. If some rankings are common to two end points, they are common along the line passing through these end points; if separate rankings appear on two end points, they have a joint value somewhere on the joining line, but this value is anyway dominated by that of common end points. Inside the limiting lines and hyperplanes, common ranking dominates.

Numerical Example (The Fighter Aircraft Problem)

Consider again the fighter aircraft problem, this time with the order of importance of the attributes given as $(X_6, X_5, X_1, X_2, X_3, X_4)$. This order can be expressed with w_j's as $w_6 \geq w_5 \geq w_1 \geq w_2 \geq w_3 \geq w_4$. The six different $\underline{w}$'s in Table 3.1 satisfy these inequality relations. The best rankings for the six different cardinal weight $\underline{w}$'s are identified in Table 3.1 and Fig. 3.3. The following decision can be drawn:

a) As long as $w_6 + w_5 > \frac{2}{3}$, A_1 is selected.

b) As soon as $w_6 + w_5 < \frac{2}{3}$, A_3 is selected.

The result matches the solution of the cardinal weight given, in which A_3 is ranked the best when $w_6 + w_5 = 0.5$.

Note

The permutation method is a useful method owing to its flexibility with regard to ordinal and cardinal rankings. A possible drawback of this method is the fact that, in the absence of a clear dominant alternative, rather complicated conditions for the values of weights may arise, particularly because numerical statements about ordinal weights are not easy to interpret [BM-18]. Also with the increase of the number of alternatives the number of permutations increases drastically.

This method measures the level of concordance of the complete preference order, whereas ELECTRE method (see Section 2.3.4) measures the concordance level of the partial order among alternatives first, then forms the possible aggregate order. Both methods can treat ordinal and cardinal preferences of attributes, but the permutation method is rather designed for ordinal cases and ELECTRE is for cardinal cases. This is the reason we classify them separately.

Table 3.1. The best order for the different $\underline{w}$'s.

$\underline{w} = (w_1, w_2, w_3, w_4, w_5, w_6)$	Best order
$(0., 0., 0., 0., 0., 1.)$	$P_3 = (A_1, A_3, A_2, A_4)$, $P_4 = (A_1, A_3, A_4, A_2)$
$(0., 0., 0., 0., 1/2, 1/2)$	$P_4 = (A_1, A_3, A_4, A_2)$, $P_{14} = (A_3, A_1, A_4, A_2)$
$(1/3, 0., 0., 0., 1/3, 1/3)$	$P_4 = (A_1, A_3, A_4, A_2)$
$(1/4, 1/4, 0., 0., 1/4, 1/4)$	$P_{10} = (A_2, A_3, A_4, A_1)$, $P_{16} = (A_3, A_2, A_4, A_1)$
$(1/5, 1/5, 1/5, 0., 1/5, 1/5)$	$P_{17} = (A_3, A_4, A_1, A_2)$
$(1/6, 1/6, 1/6, 1/6, 1/6, 1/6)$	$P_{17} = (A_3, A_4, A_1, A_2)$

Given: $w_6 \geq w_5 \geq w_1 \geq w_2 \geq w_3 \geq w_4$

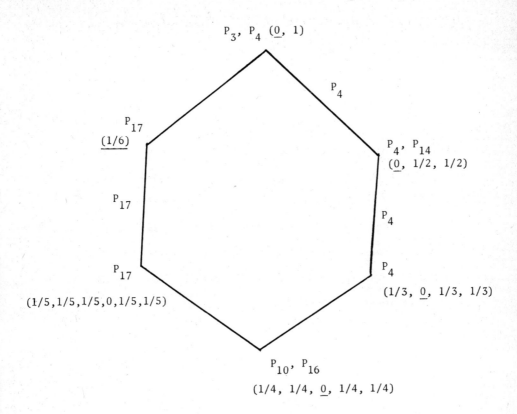

P_3, P_4 $(\underline{0},\ 1)$

P_4

P_{17}
$\underline{(1/6)}$

P_4, P_{14}
$(\underline{0},\ 1/2,\ 1/2)$

P_{17}

P_4

P_{17}

P_4

$(1/5, 1/5, 1/5, 0, 1/5, 1/5)$

$(1/3,\ \underline{0},\ 1/3,\ 1/3)$

P_{10}, P_{16}
$(1/4,\ 1/4,\ \underline{0},\ 1/4,\ 1/4)$

Figure 3.3 Graphical presentation of Table 3.1.

2.3. METHODS FOR CARDINAL PREFERENCE OF ATTRIBUTE GIVEN

The methods in this class require the DM's cardinal preferences of attributes. This (a set of weights for the attributes) is the most common way of expressing the interattribute preference information.

There are several reliable methods to assess the cardinal weights (see Section II-5 Methods for Assessing Weight).

The methods in this section all involve implicit tradeoff, but their evaluation principles are quite diverse: (1) select an alternative which has the largest utility (for simple additive weighting, hierarchical additive weighting), (2) arrange a set of overall preference rankings which best satisfies a given concordance measure (for linear assignment method and ELECTRE), and (3) select an alternative which has the largest relative closeness to the ideal solution (for TOPSIS). (See Table 1.3 for references).

2.3.1   LINEAR ASSIGNMENT METHOD

Bernardo and Blin [25] developed the linear assignment method based on a set of attributewise rankings and a set of attribute weights. The method features a linear compensatory process for attribute interaction and combination. In the process only ordinal data, rather than cardinal data, are used as the input. This information requirement is attractive in that we do not need to scale the qualitative attributes.

There is one simple way to transform the attributewise ranks into the overall ranks. The method is to pick the sum of ranks for each alternative and rank them from the lowest sum to the highest sum. For instance, consider the following attributewise preferences with equal weight,

rank	X_1	X_2	X_3
1st	A_1	A_1	A_2
2nd	A_2	A_3	A_1
3rd	A_3	A_2	A_3

Here we obtain: rank (A_1) = 1 + 1 + 2 = 4, rank (A_2) = 6, and rank (A_3) = 8. But this method fails to take full account of the compensation hypothesis, because the final rank of an alternative only depends on its own attribute-wise ranks (which are summed) without taking account of all the other alternative attributewise ranks at the same time. Thus in spite of its apparent simplicity, this method fails to meet our basic linear compensation require-ment. Required is a means of finding an overall ranking which simultaneously uses all information contained in the attributewise rankings rather than using this information sequentially as in the sum-of-ranks.

A compensatory model from this simple approach is devised. Let us define a product-attribute matrix π as a square (m x m) nonnegative matrix whose elements π_{ik} represent the frequency (or number) that A_i is ranked the kth attributewise ranking. For example, the corresponding π matrix with the equal weight on the attributes is

$$\pi = \begin{matrix} & \begin{matrix} 1st & 2nd & 3rd \end{matrix} \\ \begin{matrix} A_1 \\ A_2 \\ A_3 \end{matrix} & \begin{bmatrix} 2 & 1 & 0 \\ 1 & 1 & 1 \\ 0 & 1 & 2 \end{bmatrix} \end{matrix}$$

For the different weight $\underline{w} = (w_1, w_2, w_3) = (.2, .3, .5)$, π matrix becomes

$$\pi = \begin{bmatrix} .2+.3 & .5 & 0 \\ .5 & .2 & .3 \\ 0 & .3 & .2+.5 \end{bmatrix} = \begin{bmatrix} .5 & .5 & 0 \\ .5 & .2 & .3 \\ 0 & .3 & .7 \end{bmatrix}$$

If any attribute is tied in the ranking, for example,

rank	$X_1(w_1)$
1st	A_1, A_2
2nd	
3rd	A_3

it can be equalized by taking one more attribute with half of the original weight value such as

rank	$X_{11}(w_1/2)$	$X_{12}(w_1/2)$
1st	A_1	A_2
2nd	A_2	A_1
3rd	A_3	A_3

It is understood that π_{ik} measures the contribution of A_i to the overall ranking, if A_i is assigned to the kth overall rank. The larger π_{ik} indicates the more concordance in assigning A_i to the kth overall rank. Hence the problem is to find A_i for each k, k = 1,2,...,m which maximizes $\sum_{k=1}^{m} \pi_{ik}$. This is an m! comparison problem. An LP model is suggested for the case of large m.

Let us define permutation matrix P as (m x m) square matrix whose element P_{ik} = 1 if A_i is assigned to overall rank k, and P_{ik} = 0 otherwise. The linear assignment method can be written by the following LP format,

$$\max \sum_{i=1}^{m} \sum_{k=1}^{m} \pi_{ik} P_{ik} \tag{3.21}$$

subject to

$$\sum_{k=1}^{m} P_{ik} = 1, \qquad i = 1,2,...,m \tag{3.22}$$

$$\sum_{i=1}^{m} P_{ik} = 1, \qquad k = 1,2,...,m \tag{3.23}$$

Recall that P_{ik} = 1 if alternative i is assigned rank k, and clearly alternative i can be assigned to only one rank, therefore, we have eq. (3.22). Likewise, a given rank k can only have one alternative assigned to it; therefore, we have the constraint of eq. (3.23).

Let the optimal permutation matrix, which is the solution of the above LP problem, be P*. Then, the optimal ordering can be obtained by multiplying A by P*.

Numerical Example (The Fighter Aircraft Problem)

From the original decision matrix of the fighter aircraft problem, we can easily obtain the following attributewise preferences.

Rank	X_1	X_2	X_3	X_4	X_5	X_6
1st	A_2	A_2	A_3	A_3	A_3	A_1
2nd	A_4	A_3	A_1, A_4	A_4	A_1, A_4	A_3
3rd	A_1	A_4		A_1		A_2, A_4
4th	A_3	A_1	A_2	A_2	A_2	

Three attributes X_3, X_5, and X_6 have tied attributewise rankings. These can be equalized as

rank	X_{31}	X_{32}	X_{51}	X_{52}	X_{61}	X_{62}
1st	A_3	A_3	A_3	A_3	A_1	A_1
2nd	A_1	A_4	A_1	A_4	A_3	A_3
3rd	A_4	A_1	A_4	A_1	A_2	A_4
4th	A_2	A_2	A_2	A_2	A_4	A_2

where each of these rankings has the half weight of the tied ranking. The DM has set up the weight of $\underline{w}$ = (.2, .1, .1, .1, .2, .3), then the π matrix becomes

$$\pi = \begin{bmatrix} & 1^{st} & 2^{nd} & 3^{rd} & 4^{th} & \\ .3 & .15 & .45 & .1 & A_1 \\ .3 & 0 & .15 & .55 & A_2 \\ .4 & .4 & 0 & .20 & A_3 \\ 0 & .45 & .40 & .15 & A_4 \end{bmatrix}$$

where $\pi_{11} = w_{61} + w_{62} = 0.15 + 0.15 = 0.3$, $\pi_{12} = w_{31} + w_{51} = .05 + .1 = .15$,

$\pi_{13} = w_1 + w_{32} + w_4 + w_{52} = .2 + .05 + .1 + .1 = .45$, etc.

The LP formulation with the above π matrix is

$$\max \sum_{i=1}^{4} \sum_{k=1}^{4} \pi_{ik} P_{ik}$$

s.t.
$$\sum_{k=1}^{4} P_{ik} = 1, \qquad i = 1,2,\ldots,4$$

$$\sum_{i=1}^{4} P_{ik} = 1, \qquad k = 1,2,\ldots,4$$

$$P_{ik} \geq 0 \text{ for all } i \text{ and } k$$

The optimal permutation matrix P* is

$$P* = \begin{bmatrix} 0 & 0 & 1 & 0 \\ 0 & 0 & 0 & 1 \\ 1 & 0 & 0 & 0 \\ 0 & 1 & 0 & 0 \end{bmatrix}$$

Applying the permutation matrix P* to A, we find the optimal order:

$$A \cdot P* = A* = (A_3, A_4, A_1, A_2).$$

Note

The method gives an overall preference ranking of the alternatives based on a set of attributewise rankings and a set of attribute weights. The mathematical formulation of the model leads to a special type of linear assignment problem. For computational purposes, many efficient linear assignment computer codes are available; hence special computer packages are not necessary.

The method features a linear compensatory process for attribute interaction and combination. But only ordinal data, rather than cardinal data, are used as the input. This weaker information requirement is attractive.

Besides being able to determine the best alternative, the method has certain unique advantages in application [25]. For data collection, all that is required is the attributewise rankings. Thus we eliminate the tedious requirements of the existing compensatory models; e.g., the rather lengthy procedures of "tradeoff" analysis are not required. The procedure also eliminates the obvious difficulties encountered in constructing appropriate interval-scaled indices of attributes as required for regression analysis to be applicable. Even though a lengthy data gathering effort is eliminated, the method does satisfy the compensatory hypothesis, whereas other procedures which rely on minimal data do not. For example, the elimination by aspect (2.2.2) approach, and the lexicographic method (2.2.1) are not truly compensatory.

2.3.2   SIMPLE ADDITIVE WEIGHTING METHOD

Simple Additive Weighting method (SAW) is probably the best known and very widely used method of MADM. The method is well summarized by MacCrimmon [273]. A basic discussion of the underlying considerations is given by Churchman and Ackoff [56], and by Klee [239a]. (See Table 1.3 for references.)

To each of the attributes in SAW, the DM assigns importance weights which become the coefficients of the variables. To reflect the DM's marginal worth assessments within attributes, the DM also makes a numerical scaling of intra-attribute values. The DM can then obtain a total score for each alternative simply by multiplying the scale rating for each attribute value by the importance weight assigned to the attribute and then summing these products over all attributes. After the total scores are computed for each alternative, the alternative with the highest score (the highest weighted average) is the one prescribed to the DM.

Mathematically, simple additive weighting method can be stated as follows: Suppose the DM assigns a set of importance weights to the attributes, $\underline{w} = \{w_1, w_2, \ldots, w_n\}$. Then the most preferred alternative, A^*, is selected such that

$$A^* = \left\{A_i \ \Big| \ \max_i \ \sum_{j=1}^{n} w_j x_{ij} \Big/ \sum_{j=1}^{n} w_j \right\} \tag{3.24}$$

where x_{ij} is the outcome of the ith alternative about the jth attribute with a numerically comparable scale. Usually the weights are normalized so that $\sum_{j=1}^{n} w_j = 1$.

Numerical Example (The Fighter Aircraft Problem)

The decision matrix after assigning numerical values (by using the interval scale of Fig. 2.1) to qualitative attributes is

$$
D = \begin{array}{cccccc}
 X_1 & X_2 & X_3 & X_4 & X_5 & X_6 \\
\end{array}
$$

$$
D = \begin{bmatrix}
2.0 & 1500 & 20,000 & 5.5 & 5 & 9 \\
2.5 & 2700 & 18,000 & 6.5 & 3 & 5 \\
1.8 & 2000 & 21,000 & 4.5 & 7 & 7 \\
2.2 & 1800 & 20,000 & 5.0 & 5 & 5
\end{bmatrix}
\begin{array}{c}
A_1 \\ A_2 \\ A_3 \\ A_4
\end{array}
$$

Simple additive weighting method requires a comparable scale for all elements in the decision matrix. The comparable scale is obtained by using eq. (2.2) for benefit criteria and by eq. (2.4) for cost criteria, that is,

$$
r_{ij} = \frac{x_{ij}}{x_j^*} , \qquad j = 1,2,3,5,6
$$

and

$$
r_{i4} = \frac{x_i^{min}}{x_{i4}}
$$

The comparable numerical decision matrix is

$$
D' = \begin{array}{cccccc}
X_1 & X_2 & X_3 & X_4 & X_5 & X_6 \\
\end{array}
$$

$$
D' = \begin{bmatrix}
0.80 & 0.56 & 0.95 & 0.82 & 0.71 & 1.0 \\
1.0 & 1.0 & 0.86 & 0.69 & 0.43 & 0.56 \\
0.72 & 0.74 & 1.0 & 1.0 & 1.0 & 0.78 \\
0.88 & 0.67 & 0.95 & 0.90 & 0.71 & 0.36
\end{bmatrix}
\begin{array}{c}
A_1 \\ A_2 \\ A_3 \\ A_4
\end{array}
$$

Assume that the DM assigns the set of weights to the attributes as $\underline{w} = (0.2, 0.1, 0.1, 0.1, 0.2, 0.3)$. Then the weighted average values for the alternatives are:

$$A_1 = \sum_{j=1}^{6} w_j x_{1j} = 0.835$$

$$A_2 = 0.709$$

$$A_3 = 0.852$$

$$A_4 = 0.738$$

Therefore, A_3 is selected.

Note

Simple additive weighting method uses all n attribute values of an alternative and uses the regular arithmetical operations of multiplication and addition; therefore, the attribute values must be both numerical and comparable. Further, it is also necessary to find a reasonable basis on which to form the weights reflecting the importance of each of the attributes.

In the fighter aircraft problem, for instance, the weights, $\underline{w}$ = (0.2, 0.1, 0.1, 0.1, 0.2, 0.3), imply that maximum speed (X_1) or reliability (X_5) is twice as important as ferry range (X_2) or maximum payload (X_3) or acquisition cost (X_4). What does it mean to say that one attribute is twice as important as another? The ambiguities of such judgments must be apparent. Although techniques such as eigenvector method, weighted least square method, entropy method, etc., (see Section II-5) have been suggested for determining such weighted values, the DM must be very careful in assigning these numerical values.

The attribute values must be put into a numerical and comparable form. They must be comparable because we are going to combine across attributes; e.g., a "high" value for one attribute must receive approximately the same numerical values as "high" values of other attributes.

When weights are assigned and attribute values are numerical and comparable, some arbitrary assumptions still remain. Note that 0.36 $(=r_{46})$ multiplied by 0.3 $(=w_6)$ and 1.0 $(=r_{33})$ multiplied by 0.1 $(=w_3)$ both yield about the same product. If we interpret 1.0 as being an exceptional attribute value and

0.36 as being a below average attribute value, then this identity implies
that an exceptional maximum payload and a somewhat below average maneuverability just offset each other. By "offset each other" we mean that both
make the same contribution to the weighted average. Thus there exist some
difficulties in interpreting the output of the multiplication of attribute
values by weights.

Attributes cannot often be considered separately and then added together;
because of the complementarities between the various attributes, the approach
of weighted averages may give misleading results. But when the attributes
can in fact be considered separately (i.e., when there are essentially no
important complementarities), simple additive weighting method can be a
very powerful approach to MADM. The approach will lead to a unique choice
since a single number is arrived at for each alternative, and since these numbers
will usually be different. For this reason, and because it has some intuitive
appeal, it is frequently used [273].

The utility functions being used for uncertainty can be equally used
in the case of certainty, and is called value function $V(x_1, x_2, \ldots, x_n)$. A
value function satisfies the following property:

$$V(x_1, x_2, \ldots, x_n) \geq V(x_1', x_2', \ldots, x_n') \Leftrightarrow (x_1, x_2, \ldots, x_n) \geq (x_1', x_2', \ldots, x_n')$$

For independent attributes, a value function takes the form

$$V(x_1, x_2, \ldots, x_n) = \sum_{j=1}^{n} w_j v_j(x_j) \approx \sum_{j=1}^{n} w_j r_j$$

where $v_j(.)$ is the value function for the jth attribute and r_j is the jth
attribute transformed into the comparable scale. A utility function can be
a value function, but a value function is not necessarily a utility function;

that is,

$$U(x_1, x_2, \ldots, x_n) \Rightarrow V(x_1, x_2, \ldots, x_n)$$

Hence a valid additive utility function can be substituted for the simple additive weighting function.

Nonlinear Additive Weighting: In simple additive weighting method, it is assumed that the utility (score, value) of the multiple attributes can be separated into utilities for each of the individual attributes. When the attributes in question are complementary (that is, excellence with respect to one attribute enhances the utility of excellence with respect to another), or substitutes (that is, excellence with respect to one attribute reduces the utility gain associated with excellence with respect to other attributes), it is hard to expect that attributes take the separable additive form. Then the overall score or performance can be made in a quasi-additive or multilinear form. Basic theoretical matters are discussed by Fishburn [108, 128], and Raiffa and Keeney [BM-13]. But theory, simulation computations, and experience all suggest that simple additive weighting method yields extremely close approximations to very much more complicated nonlinear forms, while remaining far easier to use and understand [86, 90a].

2.3.3 HIERARCHICAL ADDITIVE WEIGHTING METHOD

In simple additive weighting method (SAW), the weighted averages (or priority value) for alternative A_i are given by

$$\sum_{j=1}^{n} w_j x_{ij} / \sum_{j=1}^{n} w_j$$

where in general, $\sum_{j=1}^{n} w_j = 1$, and x_{ij} is in a ratio scale.

Klee [239a] interprets the ratio x_{ij} as the subscore of the ith alternative with regard to the jth criterion. Then the vector $\underline{x}_j = (x_{1j}, x_{2j}, \ldots, x_{ij}, \ldots, x_{mj})$ may indicate the contribution, importance (another weight) of A_i's for the jth criteria as the weight vector $\underline{w}$ represents the importance of different criteria for the decision problem. If we impose

$$\sum_{i=1}^{m} x_{ij} = 1, \qquad j = 1,2,\ldots,n$$

the SAW is simple to compose weights from the different levels. This approach matches Saaty's [350] hierarchical structures. For example, the fighter air-craft decision problem by the SAW can be represented as a hierarchy with three levels. In Fig. 3.4 the first hierarchy level has a single objective, the mission effectiveness of the fighter. Its priority value is assumed to be equal to unity. The second hierarchy level has six objectives (attributes), maximum speed, ferry range, maximum payload, purchasing cost, reliability, and maneuverability. Their priorities (weights) are derived from the various weight assessing methods with respect to the objective of the first level. The third hierarchy level has the four fighters considered. In this level priorities should be derived with respect to each objective of the second level. The object is to determine the priorities of the different fighters on mission effectiveness through the intermediate second level.

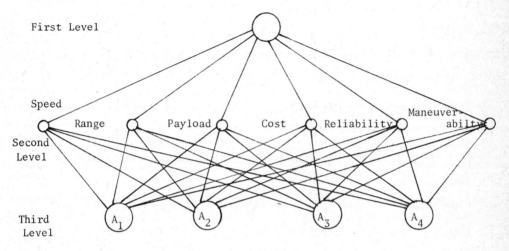

Fig. 3.4 A hierarchy for priorities of fighter aircrafts.

A formal hierarchy [350]: It is essentially a formalization in terms of partially ordered sets of our intuitive understanding of the idea. It has levels: the top level consists of a single element and each element of a given level dominates or covers (serve as a property of a purpose for) some or all of the elements in the level immediately below. The pairwise comparison matrix approach may be then applied to compare elements in a single level with respect to a purpose from the adjacent higher level. The process is repeated up the hierarchy and the problem is to compose the resulting priorities (obtained by either eigenvector method or least weighted method) in such a way as to obtain one overall priority vector of the impact of the lowest elements on the top element of the hierarchy by successive weighting and composition.

Let the symbol L_k represent the kth level of a hierarchy of h levels. Assume that $Y = \{y_1, y_2, \ldots, y_k\} \varepsilon L_k$ and that $X = \{x_1, x_2, x_i, \ldots, x_{k+1}\} \varepsilon L_{k+1}$. Also assume that there is an element $z \varepsilon L_{k-1}$ such that Y is covered by z. Then we consider the priority functions

$$w_z: \ Y \rightarrow [0, 1] \ \text{and} \ w_y: \ X \rightarrow [0, 1]$$

We construct the "priority function of the elements in X with respect to z " denoted w, w: $X \rightarrow [0, 1]$, by

$$w(x_i) = \sum_{j=1}^{k} w_{y_j}(x_i) w_z(y_j), \qquad i = 1,2,\ldots,k+1 \qquad (3.25)$$

It is obvious that this is no more than the process of weighting the influence of the element y_j on the priority of x_i by multiplying it with the importance of y_j with respect to z.

The algorithms involved will be simplified if one combines the $w_{y_j}(x_i)$ into a matrix B by setting $b_{ij} = w_{y_j}(x_i)$. If we further set $W_i = w(x_i)$ and $W'_j = w_z(y_j)$, then the above formula becomes

$$W_i = \sum_{j=1}^{k} b_{ij} W'_j, \qquad i = 1, 2, \ldots, k+1 \qquad (3.26)$$

Thus, we may speak of the priority vector W and, indeed, of the priority matrix B; this gives the final formulation W = BW'.

A hierarchy is complete if all $x \varepsilon L_k$ are dominated by every element in L_{k-1}, k = 2, ..., h. Let H be a complete hierarchy with lowest element b and h levels. Let B_k be the priority matrix of kth level, k = 1, 2, ..., h. If W' is the priority vector of the pth level with respect to some element z in the $(p-1)^{st}$ level, then the priority vector W of the qth level (p < q) with respect to z is given by

$$W = B_q B_{q-1} \cdots B_{p+1} W' \qquad (3.27)$$

Thus, the priority vector of the lowest level with respect to the element b is given by:

$$W = B_h B_{h-1} \cdots B_2 W'. \qquad (3.28)$$

If L_1 has a single element, as usual, W' is just a scalar; if more, a vector.

Numerical Example 1   (The Fighter Aircraft Problem)

The decision matrix of the fighter problem with all numerical attributes is

	X_1	X_2	X_3	X_4	X_5	X_6
A_1	2.0	1500	20000	5.5	5	9
A_2	2.5	2700	18000	6.5	3	5
A_3	1.8	2000	21000	4.5	7	7
A_4	2.2	1800	20000	5.0	5	5

D =

All elements of the fighter decision matrix are already given; thus it is not necessary to use the pairwise comparison matrix to assess the relative contribution of fighters for each criterion (attribute). This is the consistency case of matrix A in the eigenvector method. Instead priorities are directly obtained by

$$
k_j = \frac{x_{ij}}{\displaystyle\sum_{i=1}^{4} x_{ij}} \quad , \qquad j = 1,2,3,5,6
$$

$$
= \frac{1}{x_{ij}} \bigg/ \sum_{i=1}^{4} \frac{1}{x_{ij}} \quad , \qquad j = 4 \text{ (cost criterion)}
$$

Then

$$
B_3 = [p_{ij}]
$$

$$
\begin{array}{c}
 \\
= A_1 \\
A_2 \\
A_3 \\
A_4
\end{array}
\begin{array}{cccccc}
X_1 & X_2 & X_3 & X_4 & X_5 & X_6 \\
.2353 & .1875 & .2532 & .2399 & .25 & .3462 \\
.2941 & .3375 & .2278 & .2030 & .15 & .1923 \\
.2118 & .25 & .2658 & .2932 & .35 & .2692 \\
.2588 & .2250 & .2532 & .2639 & .25 & .1923
\end{array}
$$

Assume that the DM's assessment of weights about six attributes be

$$
B_2 = \underline{w} = [.2 \quad .1 \quad .1 \quad .1 \quad .2 \quad .3]^T
$$

The composite vector for the hierarchy of mission effectiveness (see Fig. 3.4) is given by

$$
W = B_3 \times B_2 \times W'
$$

$$
= \begin{array}{c}
A_1 \\
A_2 \\
A_3 \\
A_4
\end{array}
\begin{bmatrix}
.269 \\
.223 \\
.274 \\
.234
\end{bmatrix}
$$

where $W' = 1$ (because of single element in L_1). A_3 should be selected. Furthermore it is noticed that both SAW and hierarchical additive weighting give the same preference ordering, i.e., $A_3 > A_1 > A_4 > A_2$.

Numerical Example 2 (The Choosing of a Job [350])

A student who had just received his Ph.D. degree was offered three jobs. Six attributes were selected for the comparison. They are: research, growth, benefits, colleagues, location, and reputation. His criteria for selecting the jobs and their pairwise comparison matrix are given in Table 3.2. Due to the vague nature of the criteria, he constructs the pairwise comparison matrices of the jobs with respect to each criteria rather than the decision matrix. They are given in Table 3.3.

The eigenvalues of the matrix of the Table 3.2 is λ_{max} = 6.35 and the corresponding eigenvector is

$$B_2 = [\underset{RS}{.16} \quad \underset{G}{.19} \quad \underset{B}{.19} \quad \underset{C}{.05} \quad \underset{L}{.12} \quad \underset{RP}{.30}]^T$$

The eigenvalue and the eigenvectors of the remaining matrices are given by

λ_{max}	3.02	3.02	3.56	3.05	3.	3.21
	RS	G	B	C	L	RP

$$B_3 = \begin{array}{c} A \\ B \\ C \end{array} \begin{bmatrix} .14 & .10 & .32 & .28 & .47 & .77 \\ .63 & .33 & .22 & .65 & .47 & .17 \\ .24 & .57 & .46 & .07 & .07 & .05 \end{bmatrix}$$

The composite vector for the job with h=3 is given by

$$W = B_3 \times B_2 \times W'$$

$$= B_3 \times B_2$$

$$= \begin{array}{c} A \\ B \\ C \end{array} \begin{bmatrix} .40 \\ .34 \\ .26 \end{bmatrix}$$

The differences were sufficiently large for the candidate to accept the offer of job A.

Table 3.2   Overall satisfaction with job

	Research	Growth	Benefits	Colleagues	Location	Reputation
Research	1	1	1	4	1	$\frac{1}{2}$
Growth	1	1	2	4	1	$\frac{1}{2}$
Benefits	1	$\frac{1}{2}$	1	5	3	$\frac{1}{2}$
Colleagues	$\frac{1}{4}$	$\frac{1}{4}$	$\frac{1}{5}$	1	$\frac{1}{3}$	$\frac{1}{3}$
Location	1	1	$\frac{1}{3}$	3	1	1
Reputation	2	2	2	3	1	1

Table 3.3   Comparison of jobs with respect to six attributes

Research (RS)

	A	B	C
A	1	$\frac{1}{4}$	$\frac{1}{2}$
B	4	1	3
C	2	$\frac{1}{3}$	1

Growth (G)

	A	B	C
A	1	$\frac{1}{4}$	$\frac{1}{5}$
B	4	1	$\frac{1}{2}$
C	5	2	1

Benefits (B)

	A	B	C
A	1	3	$\frac{1}{3}$
B	$\frac{1}{3}$	1	1
C	3	1	1

Colleagues (C)

	A	B	C
A	1	$\frac{1}{3}$	5
B	3	1	7
C	$\frac{1}{5}$	$\frac{1}{7}$	1

Location (L)

	A	B	C
A	1	1	7
B	1	1	7
C	$\frac{1}{7}$	$\frac{1}{7}$	1

Reputation (RP)

	A	B	C
A	1	7	9
B	$\frac{1}{7}$	1	5
C	$\frac{1}{9}$	$\frac{1}{5}$	1

Numerical Example 3 (Psychotherapy [350])

The hierarchical method may be used to provide insight
into psychological problem areas in the following manner: Consider an individual's overall well-being as the single top level entry in a hierarchy.
Conceivably this level is primarily affected by childhood, adolescent, and
adult experiences. Factors in growth and maturity which impinge upon well-
being may be the influences of the father and the mother separately as well
as their influences together as parents, the socioeconomic backbround; sibling
relationships, one's peer group, schooling, religious status, and so on.

Here we consider a highly restricted form of the psychological problem,
where the individual in question feels that his self-confidence has been
severely undermined and his social adjustments have been impaired by a
restrictive situation during childhood. He is questioned about his childhood experiences only and is asked to relate the following elements pairwise on each level:

Level I   Overall well-being (O.W.)

Level II   Self-respect, sense of security, ability

             to adapt to others, (R, S, A).

Level III   Visible affection shown for subject (V),

             Ideas of strictness, ethics (E),

             Actual disciplining of child (D),

             Emphasis on personal adjustment with others (O).

Level IV   Influence of mother, father, both (M, F, B),

The actual hierarchy is shown in Fig. 3.5.

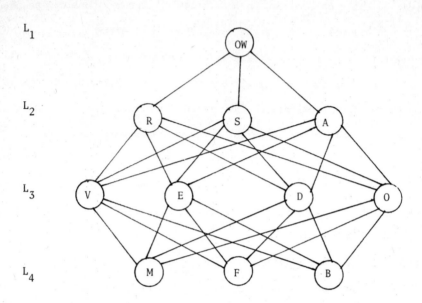

Fig. 3.5 A hierarchy of overall well-being.

The replies in the matrix form were as follows:

i) for L_2

O.W.

	R	S	A
R	1	6	4
S	$\frac{1}{6}$	1	3
A	$\frac{1}{4}$	$\frac{1}{3}$	1

ii) for L_3

R

	V	E	D	O
V	1	6	6	3
E	$\frac{1}{6}$	1	4	3
D	$\frac{1}{6}$	$\frac{1}{4}$	1	$\frac{1}{2}$
O	$\frac{1}{3}$	$\frac{1}{3}$	2	1

S

	V	E	D	O
V	1	6	6	3
E	$\frac{1}{6}$	1	4	3
D	$\frac{1}{6}$	$\frac{1}{4}$	1	$\frac{1}{2}$
O	$\frac{1}{3}$	$\frac{1}{3}$	2	1

A

	V	E	D	O
V	1	$\frac{1}{5}$	$\frac{1}{3}$	1
E	5	1	4	$\frac{1}{5}$
D	3	$\frac{1}{4}$	1	$\frac{1}{4}$
O	1	5	4	1

iii) for L_4

V

	M	F	B
M	1	9	4
F	$\frac{1}{9}$	1	8
B	$\frac{1}{4}$	$\frac{1}{8}$	1

E

	M	F	B
M	1	1	1
F	1	1	1
B	1	1	1

D

	M	F	B
M	1	9	6
F	$\frac{1}{9}$	1	$\frac{1}{4}$
B	$\frac{1}{6}$	4	1

O

	M	F	B
M	1	5	5
F	$\frac{1}{5}$	1	$\frac{1}{3}$
B	$\frac{1}{5}$	3	1

The eigenvector of the first matrix is given by

$$B_2 = \begin{matrix} R \\ S \\ A \end{matrix} \begin{bmatrix} .701 \\ .193 \\ .106 \end{bmatrix} \quad \text{O.W.}$$

The matrix of eigenvectors of the second row of matrices is given by

$$B_3 = \begin{matrix} V \\ E \\ D \\ O \end{matrix} \begin{bmatrix} .604 & .604 & .127 \\ .213 & .213 & .281 \\ .064 & .064 & .120 \\ .119 & .119 & .463 \end{bmatrix} \quad \begin{matrix} R & S & A \end{matrix}$$

The matrix of eigenvectors of the third row of matrices is given by

$$
B_4 = \begin{array}{c} \\ M \\ F \\ B \end{array}
\begin{array}{cccc} V & E & D & O \\ \left[\begin{array}{cccc} .721 & .333 & .713 & .701 \\ .210 & .333 & .061 & .097 \\ .069 & .333 & .176 & .202 \end{array}\right] \end{array}
$$

The final composite vector of influence on well-being obtained from the product $B_4 B_3 B_2$ is given by

$$
W = \begin{array}{c} M \\ F \\ B \end{array}
\left[\begin{array}{c} .635 \\ .209 \\ .156 \end{array}\right]
$$

Note

When the decision problem has a large number of attributes, say more than seven, it is easier to assess the set of weights using the hierarchical structure of the objectives. Miller [BM-17] gives a job choice example in which he identifies four major attributes influencing the choice: monetary compensation, geographical location, travel requirements, and nature of work. These factors are then decomposed into the more specific attributes characterizing each of the possible jobs. For instance there are climate, degree of urbanity, and proximity to relatives under the class of geographical location. First he assesses the weights of the four major attributes, then he judges within each class. The composition of these different weights leads to the weights on the lowest level attributes. MacCrimmon termed this approach hierarchical additive weighting in his review [275]. It is true that both Miller and Saaty use the hierarchical structure for different purposes; assessing weights, evaluating alternatives. But we feel that Saaty's approach deserves to be named "Hierarchical Additive Weighting Method" in the sense that it is the vertical extension of the SAW.

When a hierarchy has only three levels, it is equivalent to SAW. Hence Saaty's approach is really worth utilizing when $h > 3$.

2.3.4   ELECTRE METHOD

The ELECTRE method (Elimination et Choice Translating Reality) was originally introduced by Benayoun et al. [20]. Since then Roy, Nijkamp, van Delft et al. [303, 344, 345, 349, 413] have developed this method to the present state.

ELECTRE uses the concept of an 'outranking relationship'. The outranking relationship of $A_k \rightarrow A_\ell$ says that even though two alternatives k and ℓ do not dominate each other mathematically (see 1.1.1 Dominance), the DM accepts the risk of regarding A_k as almost surely better than A_ℓ [345]. Through the successive assessments of the outranking relationships of the other alternatives, the dominated alternatives defined by the outranking relationship can be eliminated. But the construction of this partial order is not an unambiguous task for the DM. ELECTRE sets the criteria for the mechanical assessment of the outranking relationships.

This method consists of a pairwise comparison of alternatives based on the degree to which evaluations of the alternatives and the preference weights confirm or contradict the pairwise dominance relationships between alternatives. It examines both the degree to which the preference weights are in agreement with pairwise dominance relationships and the degree to which weighted evaluations differ from each other. These stages are based on a 'concordance and discordance' set, hence this method is also called concordance analysis [305].

The ELECTRE method takes the following steps:

Step 1.   Calculate the normalized decision matrix; this procedure transforms the various attribute scales into comparable scales.

Each normalized value r_{ij} of the normalized decision matrix R can be calculated as:

$$r_{ij} = \frac{x_{ij}}{\sqrt{\sum\limits_{i=1}^{m} x_{ij}^2}} \quad , \qquad R = \begin{bmatrix} r_{11} & r_{12} & \cdots & r_{1n} \\ r_{21} & r_{22} & \cdots & r_{2n} \\ \cdot \\ \cdot \\ \cdot \\ r_{m1} & r_{m2} & \cdots & r_{mn} \end{bmatrix} \qquad (3.29)$$

so that all attributes have the same unit length of vector.

Step 2. Calculate the weighted normalized decision matrix: This matrix can be calculated by multiplying each column of matrix R with its associated weight w_j. Therefore, the weighted normalized decision matrix V is equal to

$$V = RW$$

$$= \begin{bmatrix} v_{11} & \cdots & v_{1j} & \cdots & v_{1n} \\ \cdot \\ \cdot \\ \cdot \\ v_{m1} & \cdots & v_{mj} & \cdots & v_{mn} \end{bmatrix} = \begin{bmatrix} w_1 r_{11} & \cdots & w_j r_{1j} & \cdots & w_n r_{1n} \\ \cdot \\ \cdot \\ \cdot \\ w_1 r_{m1} & \cdots & w_j r_{mj} & \cdots & w_n r_{mn} \end{bmatrix}$$

$$(3.30)$$

where

$$W = \begin{bmatrix} w_1 & & & & 0 \\ & w_2 \\ & & \cdot \\ & & & \cdot \\ 0 & & & & w_m \end{bmatrix}$$

Step 3. Determine the concordance and discordance set: For each pair of alternatives k and ℓ (k, ℓ = 1,2,...,m and k $\neq$ ℓ), the set of decision criteria J = $\{j | j = 1,2,...,n\}$ is divided into two distinct subsets. The concordance set $C_{k\ell}$ of A_k and A_ℓ is composed of all criteria for which A_k is preferred to A_ℓ. In other words,

$$C_{k\ell} = \{j | x_{kj} \geq x_{\ell j}\} \tag{3.31}$$

The complementary subset is called the discordance set, which is

$$D_{k\ell} = \{j | x_{kj} < x_{\ell j}\}$$

$$= J - C_{k\ell} \tag{3.32}$$

Step 4. Calculate the concordance matrix: The relative value of the concordance set is measured by means of the concordance index. The concordance index is equal to the sum of the weights associated with those criteria which are contained in the concordance set. Therefore, the concordance index $c_{k\ell}$ between A_k and A_ℓ is defined as:

$$c_{k\ell} = \sum_{j \in C_{k\ell}} w_j / \sum_{j=1}^{n} w_j$$

For the normalized weight set

$$c_{k\ell} = \sum_{j \in C_{k\ell}} w_j \tag{3.33}$$

The concordance index reflects the relative importance of A_k with respect to A_ℓ. Obviously, $0 \leq c_{k\ell} \leq 1$. A higher value of $c_{k\ell}$ indicates that A_k is preferred to A_ℓ as far as the concordance criteria are concerned.

The successive values of the concordance indices $c_{k\ell}$ (k, ℓ = 1, 2, ..., m and k $\neq$ ℓ) form the concordance matrix C of (m x m):

$$
C = \begin{bmatrix}
\text{---} & c_{12} & \cdots\cdots & c_{1m} \\
c_{21} & \text{---} & c_{23} & c_{21} \\
\vdots & & & \\
c_{m1} & c_{m2} & \cdots\cdots & c_{m(m-1)} & \text{---}
\end{bmatrix}
\tag{3.34}
$$

It should be noted that matrix C is, in general, not symmetric.

Step 5. Calculate the discordance matrix: The concordance index reflects the relative dominance of a certain alternative over a competing alternative on the basis of the relative weight attached to the successive decision criteria. So far no attention has been paid to the degree to which the evaluations of a certain A_k are worse than the evaluations of competing A_ℓ. Therefore a second index, called the discordance index, has to be defined,

$$
d_{k\ell} = \frac{\max\limits_{j \in D_{k\ell}} \left| v_{kj} - v_{\ell j} \right|}{\max\limits_{j \in J} \left| v_{kj} - v_{\ell j} \right|}
\tag{3.35}
$$

It is clear that $0 \leq d_{k\ell} \leq 1$. A higher value of $d_{k\ell}$ implies that, for the discordance criteria, A_k is less favorable than A_ℓ, and a lower value of $d_{k\ell}$, A_k is favorable to A_ℓ. The discordance indices form the discordance matrix D_x of $(m \times m)$:

$$D_x = \begin{bmatrix} \text{---} & d_{12} & \cdots & d_{1m} \\ d_{21} & \text{---} & d_{23} & d_{2m} \\ \vdots & & & \vdots \\ d_{m1} & & \cdots & d_{m(m-1)} \text{---} \end{bmatrix} \qquad (3.36)$$

Obviously, matrix D_x is, in general, asymmetric.

It should be noticed that the information contained in the concordance matrix differs significantly from that contained in the discordance matrix, making the information content C and D_x complementary; differences among weights are represented by means of the concordance matrix, whereas differences among attribute values are represented by means of the discordance matrix.

Step 6. Determine the concordance dominance matrix: This matrix can be calculated with the aid of a threshold value for the concordance index. A_k will only have a chance of dominating A_ℓ, if its corresponding concordance index $c_{k\ell}$ exceeds at least a certain threshold value $\bar{c}$, i.e.,

$$c_{k\ell} \geq \bar{c}$$

This threshold value can be determined, for example, as the average concordance index, i.e.,

$$\bar{c} = \sum_{\substack{k=1 \\ k \neq \ell}}^{m} \sum_{\substack{\ell=1 \\ \ell \neq k}}^{m} c_{k\ell}/m(m-1) \qquad (3.37)$$

On the basis of the threshold value, a Boolean matrix F can be constructed, the elements of which are defined as

$$f_{k\ell} = 1, \text{ if } c_{k\ell} \geq \bar{c} \tag{3.38}$$

$$f_{k\ell} = 0, \text{ if } c_{k\ell} < \bar{c}$$

Then each element of 1 on the matrix F represents a dominance of one alternative with respect to another one.

Step 7. Determine the discordance dominance matrix: This matrix is constructed in a way analogous to the F matrix on the basis of a threshold value $\bar{d}$ to the discordance indices. The elements of $g_{k\ell}$ of the discordance dominance matrix G are calculated as

$$\bar{d} = \sum_{\substack{k=1 \\ k \neq \ell}}^{m} \sum_{\substack{\ell=1 \\ \ell \neq k}}^{m} d_{k\ell}/m(m-1) \tag{3.39}$$

$$g_{k\ell} = 1, \text{ if } d_{k\ell} \leq \bar{d}$$

$$g_{k\ell} = 0, \text{ if } d_{k\ell} > \bar{d}$$

Also the unit elements in the G matrix represent the dominance relationships between any two alternatives.

Step 8. Determine the aggregate dominance matrix: The next step is to calculate the intersection of the concordance dominance matrix F and discordance dominance matrix G. The resulting matrix, called the aggregate dominance matrix E, is defined by means of its typical elements $e_{k\ell}$ as:

$$e_{k\ell} = f_{k\ell} \cdot g_{k\ell} \qquad\qquad (3.40)$$

Step 9. Eliminate the less favorable alternatives: The aggregate dominance matrix E gives the partial-preference ordering of the alternatives. If $e_{k\ell} = 1$, then A_k is preferred to A_ℓ for both the concordance and discordance criteria, but A_k still has the chance of being dominated by the other alternatives. Hence the condition that A_k is not dominated by ELECTRE procedure is,

$$e_{k\ell} = 1, \text{ for at least one } \ell, \; \ell = 1,2,\ldots,m, \; k \neq \ell \qquad (3.41)$$

$$e_{ik} = 0, \text{ for all } i, \; i = 1,2,\ldots,m, \; i \neq k, \; i \neq \ell$$

This condition appears difficult to apply, but the dominated alternatives can be easily identified in the E matrix. If any column of the E matrix has at least one element of 1, then this column is 'ELECTREcally' dominated by the corresponding row(s). Hence we simply eliminate any column(s) which have an element of 1.

Numerical Example (The Fighter Aircraft Problem)

Qualitative attributes are quantified with the scale of; very high = 9, high = 7, average = 5, and low = 3; then the decision matrix of the fighter problem is,

$$D = \begin{array}{c} \\ \\ A_1 \\ A_2 \\ A_3 \\ A_4 \end{array} \begin{array}{cccccc} X_1 & X_2 & X_3 & X_4 & X_5 & X_6 \\ \left[\begin{array}{cccccc} 2.0 & 1.5 & 2.0 & 5.5 & 5 & 9 \\ 2.5 & 2.7 & 1.8 & 6.5 & 3 & 5 \\ 1.8 & 2.0 & 2.1 & 4.5 & 7 & 7 \\ 2.2 & 1.8 & 2.0 & 5.0 & 5 & 5 \end{array} \right] \end{array}$$

The DM determines the preference weight of $\underline{w}$ = (.2, .1, .1, .1, .2, .3).

<u>Step 1.</u> Calculate the normalized decision matrix:

$$
R = \begin{bmatrix}
.4671 & .3662 & .5056 & .5063 & .4811 & .6708 \\
.5839 & .6591 & .4550 & .5983 & .2887 & .3727 \\
.4204 & .4882 & .5380 & .4143 & .6736 & .5217 \\
.5139 & .4392 & .5056 & .4603 & .4811 & .3727
\end{bmatrix}
$$

<u>Step 2.</u> Calculate the weighted normalized decision matrix:

$$
V = \begin{bmatrix}
.0934 & .0366 & .0506 & .0506 & .0962 & .2012 \\
.1168 & .0659 & .0455 & .0598 & .0577 & .1118 \\
.0841 & .0488 & .0531 & .0414 & .1347 & .1565 \\
.1028 & .0439 & .0506 & .0460 & .0962 & .1118
\end{bmatrix}
$$

<u>Step 3.</u> Determine the concordance and discordance sets: Remember X_4 is cost attribute, then $C_{k\ell}$ and $D_{k\ell}$ are obtained. They are:

C_{12} = {3, 4, 5, 6} D_{12} = {1, 2}

C_{13} = {1, 6} D_{13} = {2, 3, 4, 5}

C_{14} = {3, 5, 6} D_{14} = {1, 2, 4}

C_{21} = {1, 2} D_{21} = {3, 4, 5, 6}

C_{23} = {1, 2} D_{23} = {3, 4, 5, 6}

C_{24} = {1, 2, 6} D_{24} = {3, 4, 5}

C_{31} = {2, 3, 4, 5} D_{31} = {1, 6}

C_{32} = {3, 4, 5, 6} D_{32} = {1, 2}

C_{34} = {2, 3, 4, 5, 6} D_{34} = {1}

C_{41} = {1, 2, 3, 4, 5} D_{41} = {6}

C_{42} = {3, 4, 5, 6} D_{42} = {1, 2}

C_{43} = {1} D_{43} = {2, 3, 4, 5, 6}

Step 4. Calculate the concordance matrix: An element c_{12} of the C matrix

is calculated as

$$c_{12} = \sum_{j \epsilon C_{12}} w_j = w_3 + w_4 + w_5 + w_6 = .7$$

Then the concordance matrix is

$$C = \begin{bmatrix} -- & .7 & .5 & .6 \\ .3 & -- & .3 & .6 \\ .5 & .7 & -- & .8 \\ .7 & .7 & .2 & -- \end{bmatrix}$$

Step 5. Calculate the discordance matrix: An element d_{12} of the D_x matrix

is obtained as

$$d_{12} = \frac{\max_{j \epsilon D_{12}} |v_{1j} - v_{2j}|}{\max_{j \epsilon J} |v_{1j} - v_{2j}|}$$

$$= \frac{\max\{.0234, .0293\}}{\max\{.0234, .0293, .0051, .0092, .0385, .0894\}}$$

$$= \frac{.0293}{.0894} = .3277$$

The discordance matrix is

$$D_x = \begin{bmatrix} -- & .3277 & .8613 & .1051 \\ 1. & -- & 1. & 1. \\ 1. & .4247 & -- & .4183 \\ 1. & .5714 & 1. & -- \end{bmatrix}$$

Step 6. Determine the concordance dominance matrix: If we take the threshold value of $c_{k\ell}$ as the average concordance index, then

$$\bar{c} = \frac{\sum\limits_{k=1}^{4} \sum\limits_{\ell=1}^{4} c_{k\ell}}{4 \times 3} = .55$$

The concordance dominance matrix is

$$F = \begin{bmatrix} \text{---} & 1 & 0 & 1 \\ 0 & \text{---} & 0 & 1 \\ 0 & 1 & \text{---} & 1 \\ 1 & 1 & 0 & \text{---} \end{bmatrix}$$

Step 7. Determine the discordance dominance matrix: The value of d is taken as the average discordance index, then

$$\bar{d} = \frac{\sum\limits_{k=1}^{4} \sum\limits_{\ell=1}^{4} d_{k\ell}}{4 \times 3} = .7257$$

Then the discordance dominance matrix is

$$G = \begin{bmatrix} \text{---} & 1 & 0 & 1 \\ 0 & \text{---} & 0 & 0 \\ 0 & 1 & \text{---} & 1 \\ 0 & 1 & 0 & \text{---} \end{bmatrix}$$

Step 8.   Determine the aggregate dominance matrix: Combining matrices of F

and G, the aggregate dominance matrix is obtained as

$$E = \begin{bmatrix} -\!- & 1 & 0 & 1 \\ 0 & -\!- & 0 & 0 \\ 0 & 1 & -\!- & 1 \\ 0 & 1 & 0 & -\!- \end{bmatrix}$$

Step 9.   Eliminate the less favorable alternatives: The E matrix renders

the following overranking relationships: $A_1 \rightarrow A_2$, $A_1 \rightarrow A_4$, $A_3 \rightarrow A_2$, $A_3 \rightarrow A_4$,

$A_4 \rightarrow A_2$.   These relationships are well illustrated by the graphical represen-

tation,

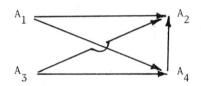

We can easily see that A_2 and A_4 are dominated by A_1 and A_3.   But we cannot

tell the preference relation between A_1 and A_3.   This relationship can also

be directly identified from the E matrix (see the first and the third column

of the E matrix which do not have any element of 1.)   Hence A_2 and A_4 can be

eliminated by ELECTRE method.

Note

A weak point of the ELECTRE method is the use of threshold values $\bar{c}$

and $\bar{d}$.   These values are rather arbitrary, although their impact on the

final solution may be significant.   For example , if we take a threshold value of

$\bar{c} = 1$, and $\bar{d} = 0$ for   complete dominance, then it is rather difficult to

eliminate any of the alternatives (in the aircraft problem no alternative

can be discarded).   And by relaxing the threshold value ($\bar{c}$ lowered; $\bar{d}$ in-

creased) we can reduce the number of nondominated solutions to the single one (in the aircraft problem if $\bar{d}$ = .9 is taken instead of the average dis-

cordance index of .7257, then only A_1 is left as a nondominated alternative).

van Delft and Nijkamp [BM-18] introduced the net dominance relation-

ships for the complementary analysis of the ELECTRE method. First they define

the net concordance dominance value c_k, which measured the degree to which the

total dominance of the A_k exceeds the degree to which all competing alter-

natives dominate A_k, i.e.,

$$c_k = \sum_{\substack{\ell=1 \\ \ell \neq k}}^{m} c_{k\ell} - \sum_{\substack{\ell=1 \\ \ell \neq k}}^{m} c_{\ell k} \tag{3.42}$$

Similarly, the net discordance dominance value d_k is defined as

$$d_k = \sum_{\substack{\ell=1 \\ \ell \neq k}}^{m} d_{k\ell} - \sum_{\substack{\ell=1 \\ \ell \neq k}}^{m} d_{\ell k} \tag{3.43}$$

Obviously A_k has a higher chance of being accepted with the higher c_k and

the lower d_k. Hence the final selection should satisfy the condition that

its net concordance dominance value should be at a maximum and its net dis-

cordance dominance at a minimum. If one of these conditions is not satis-

fied, a certain trade off between the values of c_k and d_k has to be carried

out. The procedure is to rank the alternatives according to their net con-

cordance and discordance dominance values. The alternative that scores on

the average as the highest one can be selected as the final solution. For

the aircraft problem, the net concordance dominance values are: c_1 = .3,

c_2 = -.9, c_3 = 1., c_4 = -.4; and the net discordance dominance values are:

d_1 = -1.7059, d_2 = 1.6762, d_3 = -1.0183, d_4 = 1.048. It happens that A_3

ranks the first and A_1 the second in c_k, and that A_1 ranks first and A_3 second in d_k. Hence both A_1 and A_3 are judged acceptable for the final solution.

We consider the ELECTRE method to be one of the best methods because of its simple logic, full utilization of information contained in the decision matrix, and refined computational procedure.

2.3.5 TECHNIQUE FOR ORDER PREFERENCE BY SIMILARITY TO IDEAL SOLUTION (TOPSIS)

Yoon and Hwang [440a, 440b] developed the Technique for Order Preference by Similarity to Ideal Solution (TOPSIS) based upon the concept that the chosen alternative should have the shortest distance from the ideal solution and the farthest from the negative-ideal solution.

Assume that each attribute takes the monotonically increasing (or decreasing) utility; then it is easy to locate the "ideal" solution which is composed of all best attribute values attainable, and the "negative-ideal" solution composed of all worst attribute values attainable. One approach is to take an alternative which has the (weighted) minimum Euclidean distance to the ideal solution in a geometrical sense [381, 459]. It is argued that this alternative should be farthest from the negative-ideal solution at the same time. Sometimes the chosen alternative, which has the minimum Euclidean distance from the ideal solution, has the shorter distance (to the negative-ideal) than the other alternative(s). For example, in Fig. 3.6 an alternative A_1 has shorter distances (both to ideal solution A* and to the negative-ideal solution A⁻) than the other alternative A_2. Then it is very difficult to justify the selection of A_1. TOPSIS considers the distances to both the ideal and the negative-ideal solutions simultaneously by taking the relative closeness to the ideal solution. Dasarathy [66] used this similarity measure in clustering multidimensional data arrays. This method is simple and yields an indisputable preference order of solution.

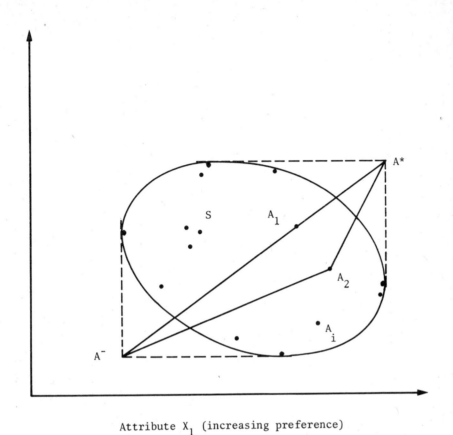

Fig. 3.6 Euclidean distances to the ideal and negative-ideal solutions in two dimensional space.

The Algorithm

The TOPSIS method evaluates the following decision matrix which contains m alternatives associated with n attributes (or criteria):

$$
D = \begin{array}{c}
\\
A_1 \\
A_2 \\
\vdots \\
A_i \\
\vdots \\
A_m
\end{array}
\begin{array}{c}
\begin{array}{cccccc}
X_1 & X_2 & & X_j & & X_n
\end{array} \\
\left[
\begin{array}{cccccc}
x_{11} & x_{12} & \cdots & x_{1j} & \cdots & x_{1n} \\
x_{21} & x_{22} & \cdots & x_{2j} & \cdots & x_{2n} \\
\vdots & \vdots & & \vdots & & \vdots \\
x_{i1} & x_{i2} & \cdots & x_{ij} & \cdots & x_{in} \\
\vdots & \vdots & & \vdots & & \vdots \\
x_{m1} & x_{m2} & \cdots & x_{mj} & \cdots & x_{mn}
\end{array}
\right]
\end{array}
$$

where

A_i = the ith alternative considered,

x_{ij} = the numerical outcome of the ith alternative with respect to the jth criterion.

TOPSIS assumes that each attribute in the decision matrix takes either monotonically increasing or monotonically decreasing utility. In other words, the larger the attribute outcomes, the greater the preference for the "benefit" criteria and the less the preference for the "cost" criteria. Further, any outcome which is expressed in a nonnumerical way should be quantified through the appropriate scaling technique. Since all criteria cannot be assumed to be of equal importance, the method receives a set of weights from the decision maker. For the sake of simplicity the proposed method will be presented as a series of successive steps.

Step 1. Construct the normalized decision matrix: This process tries to transform the various attribute dimensions into   nondimensional attributes, which allows comparison across the attributes. One way is to take the outcome of each criterion divided by the norm of the total outcome vector of the criterion at hand. An element r_{ij} of the normalized decision matrix R can be calculated as

$$r_{ij} = x_{ij} \Big/ \sqrt{\sum_{i=1}^{m} x_{ij}^2}$$

Consequently, each attribute has the same unit length of vector.

Step 2. Construct the weighted normalized decision matrix: A set of weights $\underline{w} = (w_1, w_2, \ldots, w_j, \ldots, w_n)$, $\sum_{j=1}^{n} w_j = 1$, from the decision maker is accommodated to the decision matrix in this step. This matrix can be calculated by multiplying each column of the matrix R with its associated weight w_j. Therefore, the weighted normalized decision matrix V is equal to

$$V = \begin{bmatrix} v_{11} & v_{12} & \cdots & v_{1j} & \cdots & v_{1n} \\ \vdots & \vdots & & \vdots & & \vdots \\ v_{i1} & v_{i2} & \cdots & v_{ij} & \cdots & v_{in} \\ \vdots & \vdots & & \vdots & & \vdots \\ v_{m1} & v_{m2} & \cdots & v_{mj} & \cdots & v_{mn} \end{bmatrix} = \begin{bmatrix} w_1 r_{11} & w_2 r_{12} & \cdots & w_j r_{ij} & \cdots & w_n r_{1n} \\ \vdots & \vdots & & \vdots & & \vdots \\ w_1 r_{i1} & w_2 r_{i2} & \cdots & w_j r_{ij} & \cdots & w_n r_{in} \\ \vdots & \vdots & & \vdots & & \vdots \\ w_1 r_{m1} & w_2 r_{m2} & \cdots & w_j r_{mj} & \cdots & w_n r_{mn} \end{bmatrix}$$

Step 3. Determine ideal and negative- ideal solutions: Let the two artificial alternatives A^* and A^- be defined as

$$A^* = \{ (\max_i v_{ij} \,|\, j \in J), (\min_i v_{ij} \,|\, j \in J') \,|\, i = 1, 2, \ldots, m \}$$

$$= \{ v_1^*, v_2^*, \ldots, v_j^*, \ldots, v_n^* \} \tag{3.44}$$

$$A^- = \{ (\min_i v_{ij} \,|\, j \in J), (\max_i v_{ij} \,|\, j \in J') \,|\, i = 1, 2, \ldots, m \}$$

$$= \{ v_1^-, v_2^-, \ldots, v_j^-, \ldots, v_n^- \} \tag{3.45}$$

where $J = \{j = 1,2,\ldots,n | j$ associated with benefit criteria$\}$

$J' = \{j = 1,2,\ldots,n | j$ associated with cost criteria$\}$

Then it is certain that the two created alternatives A^* and A^- indicate the most preferable alternative (ideal solution) and the least preferable alternative (negative-ideal solution), respectively.

4. <u>Calculate the separation measure</u>: The separation between each alternative can be measured by the n-dimensional Euclidean distance. The separation of each alternative from the ideal one is then given by

$$S_{i*} = \sqrt{\sum_{j=1}^{n} (v_{ij} - v_j^*)^2}, \qquad i = 1,2,\ldots,m \qquad (3.46)$$

Similarly, the separation from the negative-ideal one is given by

$$S_{i-} = \sqrt{\sum_{j=1}^{n} (v_{ij} - v_j^-)^2}, \qquad i = 1,2,\ldots,m \qquad (3.47)$$

5. <u>Calculate the relative closeness to the ideal solution</u>: The relative closeness of A_i with respect to A^* is defined as

$$C_{i*} = S_{i-}/(S_{i*} + S_{i-}), \qquad 0 < C_{i*} < 1, \qquad i = 1,2,\ldots,m \qquad (3.48)$$

It is clear that $C_{i*} = 1$ if $A_i = A^*$ and $C_{i*} = 0$ if $A_i = A^-$. An alternative A_i is closer to A^* as C_{i*} approaches to 1.

6. <u>Rank the preference order</u>: A set of alternatives can now be preference ranked according to the descending order of C_{i*}.

Numerical Example (The Fighter Aircraft Decision Problem)

The decision matrix of the fighter aircraft problem after the quantification of nonnumerical attributes of X_5 and X_6 is:

$$D = \begin{array}{c} \\ A_1 \\ A_2 \\ A_3 \\ A_4 \end{array} \begin{array}{cccccc} X_1 & X_2 & X_3 & X_4 & X_5 & X_6 \\ \left[\begin{array}{cccccc} 2.0 & 1500 & 20000 & 5.5 & 5 & 9 \\ 2.5 & 2700 & 18000 & 6.5 & 3 & 5 \\ 1.8 & 2000 & 21000 & 4.5 & 7 & 7 \\ 2.2 & 1800 & 20000 & 5.0 & 5 & 5 \end{array}\right] \end{array}$$

Note that all attributes except X_4 are the benefit criteria.

1. Calculate the normalized decision matrix

$$R = \begin{bmatrix} .4671 & .3662 & .5056 & .5063 & .4811 & .6708 \\ .5839 & .6591 & .4550 & .5983 & .2887 & .3727 \\ .4204 & .4882 & .5308 & .4143 & .6736 & .5217 \\ .5139 & .4392 & .5056 & .4603 & .4811 & .3727 \end{bmatrix}$$

2. Calculate the weighted decision matrix: Assume that the relative importance of attributes is given by the decision maker as $\underline{w} = (w_1, w_2, w_3, \ldots, w_6) = (.2, .1, .1, .1, .2, .3)$. The weighted decision matrix is then

$$V = \begin{bmatrix} .0934 & .0366 & .0506 & .0506 & .0962 & .2012 \\ .1168 & .0659 & .0455 & .0598 & .0577 & .1118 \\ .0841 & .0488 & .0531 & .0414 & .1347 & .1565 \\ .1028 & .0439 & .0506 & .0460 & .0962 & .1118 \end{bmatrix}$$

3. Determine the ideal and negative-ideal solutions:

$$A^* = (\max_i v_{i1}, \max_i v_{i2}, \max_i v_{i3}, \min_i v_{i4}, \max_i v_{i5}, \max_i v_{i6})$$

$$= (\ .1168, \quad .0659, \quad .0531, \quad .0414, \quad .1347, \quad .2012\ )$$

$$A^- = (\min_i v_{i1}, \ \min_i v_{i2}, \ \min_i v_{i3}, \ \max_i v_{i4}, \ \min_i v_{i5}, \ \min_i v_{i6})$$

$$= (\ .0841, \quad .0366, \quad .0455, \quad .0598, \quad .0577, \quad .1118\ )$$

4. Calculate the separation measures:

$$S_{i*} = \sqrt{\sum_{j=1}^{6} (v_{ij} - v_j^*)^2}, \qquad i = 1,2,3,4$$

$$S_{1*} = .0545 \qquad S_{2*} = .1197$$

$$S_{3*} = .0580 \qquad S_{4*} = .1009$$

$$S_{i-} = \sqrt{\sum_{j=1}^{6} (v_{ij} - v_j^-)^2}, \qquad i = 1,2,3,4$$

$$S_{1-} = .0983 \qquad S_{2-} = .0439$$

$$S_{3-} = .0920 \qquad S_{4-} = .0458$$

5. Calculate the relative closeness to the ideal solution:

$$C_{1*} = S_{1-}/(S_{1*} + S_{1-}) = .643, \quad C_{2*} = .268,$$
$$C_{3*} = .613, \quad C_{4*} = .312$$

6. Rank the preference order: According to the descending order of C_{i*}, the preference order is:

$$A_1, \ A_3, \ A_4, \ A_2$$

Note

Probably the best known and widely used MADM method is the Simple Additive Weighting (SAW) method (2.3.2). This method is so simple that some decision makers are reluctant to accept the solution. In this section the SAW method is reexamined through the concept of TOPSIS.

SAW chooses an alternative which has the maximum weighted average outcome. That is, to select A^+ such that

$$A^+ = \{A_i \mid \max_i \sum_{j=1}^n w_j r_{ij} / \sum_{j=1}^n w_j\}$$

where $\sum_{j=1}^n w_j = 1$ and r_{ij} is the normalized outcome of A_i with respect to the jth benefit criterion (cost criterion is converted to the benefit by taking the reciprocal before normalization). The chosen alternative A^+ can be rewritten by

$$A^+ = \{A_i \mid \max_i \sum_{j=1}^n v_{ij}\}$$

Let the separation measure in TOPSIS be defined by the city block distance [66] instead of the Euclidean distance; then the separation between A_i and A_k can be written by

$$S_{ik} = \sum_{j=1}^n | v_{ij} - v_{kj} |, \qquad i,k = 1,2,\ldots,m$$
$$i \neq k$$

This city block distance measure has the following useful relationship [440a]:

$$S_{i*} + S_{i-} = S_{*-} = K, \qquad i = 1,2,\ldots,m$$

where K is a positive constant.

The relationship says that any alternative which has the shortest distance to the ideal solution is guaranteed to have the longest distance to the negative-ideal solution. (This is not true for the Euclidean distance measure.) Now the relative closeness to the ideal solution can be simplified as

$$C_{i*} = S_{i-} \, / \, S_{*-}, \qquad i = 1,2,\ldots,m$$

Assume that the chosen alternative A^+ can be described as

$$A^+ = \{A_i | \max_i C_{i*}\}$$

It is also proved that [440a]:

$$A^+ = \{A_i \mid \max_i \sum_{j=1}^{n} v_{ij}\} = \{A_i | \max_i C_{i*}\}$$

Now it can be concluded that the result of SAW is a special case of TOPSIS using the city block distance.

The C_{i*} values of the illustrative example with the different separation measures are given below:

	Euclidean distance	City block distance
A_1	.643	.596
A_2	.268	.244
A_3	.614	.629
A_4	.312	.328

Even though they have similar values, the best alternative for the Euclidean distance measure is A_1, and that for the city block distance is A_3. The preference orders are (A_1, A_3, A_4, A_2) for the Euclidean distance measure, and (A_3, A_1, A_4, A_2) for the city block distance measure.

TOPSIS takes the cardinal preference information on attributes; that is, a set of weights for the attributes is required. The solution depends upon the weighting scheme given by the DM. Fortunately some reliable methods for weight assessment have appeared [54, 350, 459] which will enhance the utilization of the proposed method. (See Section II-5).

TOPSIS assumes that each attribute takes either monotonically increasing or decreasing utility. This monotonicity requirement is a very reasonable assumption [BM-13]. A nonmonotonic utility (e.g., square utility) is a rare case, such as the optimum number of rooms in a house or the blood sugar count in a human body, where the best utility is located somewhere in the middle of an attribute range.

The two different separation measures of TOPSIS (by the city block distance and Euclidean distance) were introduced in the previous sections. Here they are contrasted with the concept of tradeoffs [BM-13, 277, 278]. A tradeoff is a ratio of the change in one attribute that exactly offsets a change in another attribute. Indifference curves are contours on a given value (or utility) function. A DM is assumed to give equal preference value to any alternative located on an indifference curve. The tradeoff or marginal rate of substitution (MRS) at any point is the negative reciprocal of the slope at that point. Thus, if indifference curves are given, the MRS can be calculated. Mathematically if an indifference curve passing through a point (v_1, v_2) is given by

$$f(v_1, v_2) = c \qquad (3.49)$$

where f is a value function and c is a constant, then the MRS, λ, at (v_1, v_2) can be obtained, i.e.,

$$\lambda = - \frac{dv_1}{dv_2}\bigg|_{(v_1,v_2)} = \frac{\partial f}{\partial v_2}\bigg/\frac{\partial f}{\partial v_1}\bigg|_{(v_1,v_2)} \tag{3.50}$$

SAW or TOPSIS with city block distance measure has the value function of

$$f(v_1,v_2) = v_1 + v_2 \tag{3.51}$$

The MRS is then given by $\lambda = 1$ (actually $\lambda = w_2/w_1$ in X_1 and X_2 space). This implies that the MRS in SAW is constant between attributes, and the indifference curves form straight lines with the slope of -1. A constant MRS is a special rare case of MRS, which implies the local MRS is also the global MRS.

TOPSIS with Euclidean distance measure has the value function of

$$f(v_1, v_2) = \frac{S_{i-}}{S_{i-} + S_{i*}} = c$$

$$= \frac{\sqrt{(v_1 - v_1^-)^2 + (v_2 - v_2^-)^2}}{\sqrt{(v_1 - v_1^-)^2 + (v_2 - v_2^-)^2} + \sqrt{(v_1 - v_1^*)^2 + (v_2 - v_2^*)^2}} \tag{3.52}$$

The MRS is now calculated by

$$\lambda = \frac{S_{i-}^2 (v_2^* - v_2) + S_{i*}^2 (v_2 - v_2^-)}{S_{i-}^2 (v_1^* - v_1) + S_{i*}^2 (v_1 - v_1^-)} \tag{3.53}$$

It is certain that λ depends on the levels of v_1 and v_2 except at the point where distances to the ideal and negative-ideal solution are equal. When the distances are equal, i.e.,

$$\lambda = \frac{v_2^* - v_2^-}{v_1^* - v_1^-} \tag{3.54}$$

when $S_{i*} = S_{i-}$. It is not easy to illustrate the general shapes of the indifference curves in this case. The value function can be rewritten by

$$cS_{i*} - (1 - c) S_{i-} = 0 \tag{3.55}$$

where $0 < c < 1$. Then this expression indicates a variation of hyperbola where the difference of its weighted (c and (1-c)) distances from two fixed points (ideal and negative-ideal solution) is zero.

Some typical indifference curves are shown in Fig. 3.7. Any curves with $C_{i*} \geq 0.5$ are convex to the preference origin, which indicate the property of the diminishing MRS observed in most indifference curves [277], whereas indifference curves with $C_{i*} \leq .5$ are concave to the preference origin. This is an unusual case of indifference curves, but it may be interpreted as a risk-prone attitude resulting from a pessimistic situation; when a DM recognizes his solution is closer to the negative-ideal than to the ideal one, he is inclined to take one which has the best attribute with the other worst attribute rather than to take one which has two worse attributes. Hence this approach can be viewed as an amalgamation of optimistic and pessimistic decision methods which is presented by the Hurwirtz rule [178a].

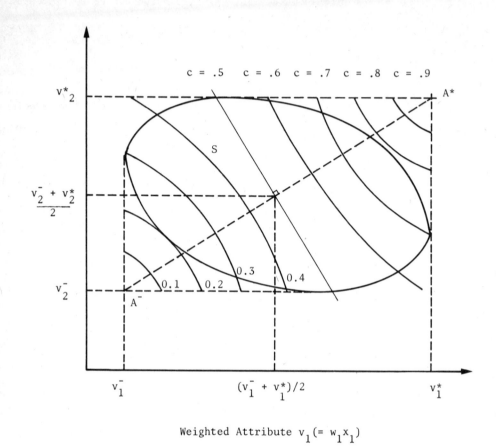

Fig. 3.7 Typical indifference curves observed in TOPSIS.

2.4   METHODS FOR MARGINAL RATE OF SUBSTITUTION OF ATTRIBUTES GIVEN

We sometimes trade in an old car plus some amount of money for a new car based upon our acceptance of the dealer's offer. The procedure for this commercial transaction is applied in multiple attribute decision making situations. If we can settle for a lower value on one attribute (e.g., reduce an amount in one's savings account), how much can we expect to get the improved value of another attribute (e.g., enjoying a new car)? Another specific example in choosing a car is that if one is willing to lower the gas mileage value, how much roomier space can one get if other things remain the same?

Most MADM methods except the noncompensatory model deal with tradeoffs implicitly or explicitly. Here we discuss a method where tradeoff information is explicitly utilized. Marginal rate of substitution (MRS) and in-difference curve are the two basic terms describing the tradeoff information.

The Marginal Rate of Substitution: Suppose that in a car selection problem, where two attributes X_1 (gas mileage) and X_2 (roominess) are specified desirable attributes while other attributes remain equal, you are asked: If X_2 is increased by Δ units, how much does X_1 have to decrease in order for you to remain indifferent? Clearly, in many cases, your answer depends on the levels x_1 of X_1 and x_2 of X_2. If, at a point (x_1, x_2), you are willing to give up $\lambda\Delta$ units of X_1 for Δ units of X_2, then we will say that the marginal rate of substitution (MRS) of X_1 for X_2 at (x_1, x_2) is λ. In other words, λ is the amount of X_1 you are willing to pay for a unit of X_2 given that you presently have x_1 of X_1 and x_2 of X_2 (see Fig. 3.8).

Making tradeoffs among three attributes is usually more difficult than making tradeoffs between two attributes. Hence we usually consider two attributes at a time.

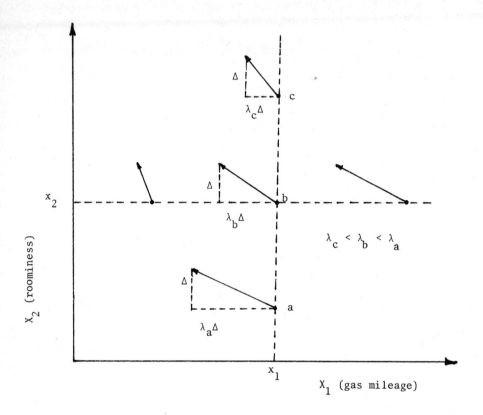

Fig. 3.8   The marginal rate of substitution as a function of X_1 and X_2.

It should be noted that when two attributes are independent of each other (noncompensatory), tradeoffs between these attributes are not relevant. In this case it is not possible to get a higher value on one attribute even though we are willing to give up a great deal of another attribute.

The marginal rate of substitution usually depends on the levels of X_1 and X_2, that is, on (x_1, x_2). For example, suppose the substitution rate at (x_1, x_2), the point b in Fig. 3.8, is λ_b. If we hold x_1 fixed, we might find that the substitution rates increase with a decrease in X_2 and decrease with an increase in X_2 as shown at points a and c in Fig. 3.8 for X_1 (gas mileage) and X_2 (roominess). The changes in the substitution rates mean that the more of X_2 (roominess) we have, the less of X_1 (gas mileage) we would be willing to give up to gain additional amounts of X_2, and the sacrifice of X_1 is less at point c than at point a. This implies that the MRS at which you would give up X_1 for gaining X_2 decreases as the level of X_2 increases, i.e., the MRS diminishes.

Indifference Curves: Let us consider again the car selection problem with X_1 (gas mileage) and X_2 (roominess expressed by the space in the passenger compartment). Consider A_1 as a reference point, a car whose gas mileage is 26 MPG and whose passenger compartment is 81 cu ft. A_1 can be expressed, then, as (26, 81) in Fig. 3.9. The indifference curve would require a new alternative, say A_2 (20, 95), that the DM would deem equivalent in preference to A_1. By obtaining a number of such points, we would trace out a curve of indifference through A_1. The indifference curve is, then, the locus of all attribute values indifferent to a reference point. We can draw any number of indifference curves with the different reference points.

The indifference curve is particularly useful because it divides the set of all attribute values into (a) those indifferent to the reference point, (b) those preferred to the reference point, and (c) those to which the reference

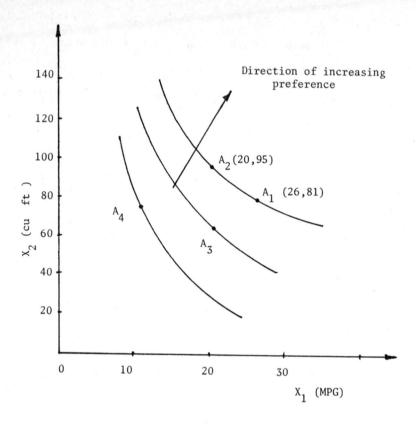

Fig. 3.9   A set of indifference curves between MPG and the space
of passenger compartment.

point is preferred. We know that any point on the preferred side of the indifference curve is preferred to any point on the curve or on the nonpreferred side of the curve. Hence, if we are asked to compare A_1 with any other alternative, we can immediately indicate a choice. However, if we are given several points on the preferred side of the indifference curve, we cannot say which is the most preferred. It would be necessary to draw new indifference curves [278]. See the preference relationships of $A_1 \sim A_2 > A_3 > A_4$ in Fig. 3.9.

Three major properties are assumed for indifference curves [277]. One such property is nonintersection. Intersection implies an intransitivity of preference and its occurrence would generally indicate a hasty consideration of preferences. A second property relates to the desirability of the attributes considered. If we assume both attributes are desirable, then in order to get more of one attribute we would be willing to give up some amount of a second attribute. This leads to a negatively sloped curve to the preference origin (not to the mathematical origin). A third property is an empirical matter. The indifference curves are assumed to be convex to the preference origin. This implies that the MRS diminishes. Note that the slope of indifference curves in Fig. 3.9 gets steeper as we move up the curves (from right to left). Also it is observed that the MRS at (x_1, x_2) is the negative reciprocal of the slope of the indifference curve at (x_1, x_2). Thus if we have indifference curves, then we can directly calculate the MRS.

MacCrimmon et al. [276, 277, 278] suggest some effective methods for obtaining indifference curves. One of their methods has the three types of structured procedures: (a) generating points by fixing only one attribute; (b) generating points by fixing both attributes, but fixing them one at a time; and (c) generating points by fixing both attributes simultaneously. Keeney and Raiffa [BM-13] introduce a method using the MRS information actively.

2.4.1 HIERARCHICAL TRADEOFFS

When interdependency exists among attributes, the consideration of trade-offs allows us to make the alternatives much more comparable than they are initially. That is, we can make alternatives equivalent for all attributes except one by tradeoffs, and then evaluate the alternatives by the attribute values of the remaining one [273, 278].

The simplest way to deal with tradeoffs on n attributes is to ignore all but two attributes; then we discard attributes one by one through the trade-offs between the natural combination of two attributes. The indifference curves easily facilitate this equivalization process. Suppose alternatives are located on the indifference curves. We can easily drive the one attribute level to the same, and read the corresponding modified value of the other attribute. The attribute which is driven to the same level (called base level) is no longer necessary for further consideration. If this procedure can be carried through for pairs of the remaining n-2 attributes, we will have a new set of n/2 attributes. Similarly, if these composite attributes also have pairs of natural combinations we can consider the tradeoffs among the pairs and use the indifference curves we obtain to scale a new higher order composite attribute. We can continue this hierarchical combination until we obtain two high order composite attributes for which we again form the tradeoff. As a result, we will incorporate all the attributes.

To select the preferred alternative with this approach, we must be able to locate it in the final composite space. This can be done by assuring that each alternative is on an indifference curve in the initial spaces; thus, the combination of values defining an alternative will be one of the scale values for the new attribute. By including these combinations on an indifference curve each step of the way, we can ensure that the alternatives will be representable in the highest order space we finally consider.

The use of this method requires that the attributes be independent among the initial classes. That is, while the tradeoff between any initial pair can be nonconstant and highly interrelated, this tradeoff cannot depend on the level of the other attributes. This restriction suggests that a useful way to form the initial pairs is by grouping attributes that seem relatively independent from the other ones.

Numerical Example (The fighter aircraft problem)

The decision matrix of the fighter aircraft problem is

$$D = \begin{array}{cccccc} X_1 & X_2 & X_3 & X_4 & X_5 & X_6 \\ \begin{bmatrix} 2.0 & 1500 & 20000 & 5.5 & \text{average} & \text{very high} \\ 2.5 & 2100 & 18000 & 6.5 & \text{low} & \text{average} \\ 1.8 & 2000 & 21000 & 4.5 & \text{high} & \text{high} \\ 2.2 & 1800 & 20000 & 5.0 & \text{average} & \text{average} \end{bmatrix} & \begin{array}{c} A_1 \\ A_2 \\ A_3 \\ A_4 \end{array} \end{array}$$

We assume that the following three pairs of attributes are independent of each other and that the interdependency exists within each pair; speed (X_1) vs reliability (X_5), range (X_2) vs payload (X_3), and cost (X_4) vs maneuverability (X_6).

First we draw a set of indifference curves for X_1 vs X_5 in Fig. 3.10. If we take 'average' value of reliability as the base level, then the following equivalization of X_5 can be obtained from Fig. 3.10.

A_1 = (2. Mach, average) ~ (2. Mach, average)

A_2 = (2.5 Mach, low) ~ (1.875 Mach, average)

A_3 = (1.8 Mach, high) ~ (2.6 Mach, average)

A_4 = (2.2 Mach, average) ~ (2.2 Mach, average)

The above modified value of maximum speed will represent both X_1 and X_5.

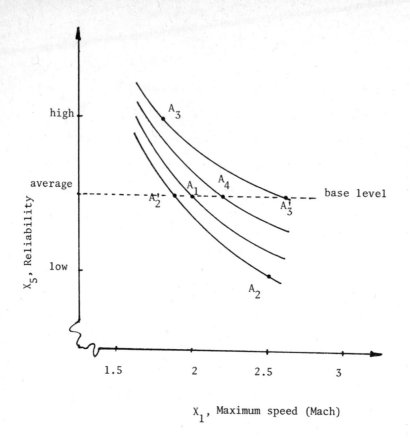

Fig. 3.10   A set of indifference curves between maximum speed and reliability.

The indifference curves for X_2 vs X_3 are given in Fig. 3.11. The following tradeoffs are made by taking maximum payload of 20,000 pounds as the base level:

A_1 = (1500 NM, 20000 pounds) ~ (1500 NM, 20000 pounds)

A_2 = (2700 NM, 18000 pounds) ~ (2200 NM, 20000 pounds)

A_3 = (2000 NM, 21000 pounds) ~ (2200 NM, 20000 pounds)

A_4 = (1800 NM, 20000 pounds) ~ (1800 NM, 20000 pounds)

Thirdly a set of indifference curves for X_4 vs X_6 are given in Fig. 3.12. The following tradeoffs are drawn by taking the 'high' level of maneuverability as the base level:

A_1 = ($5.5 million, very high) ~ ($4.5 million, high)

A_2 = ($6.5 million, average) ~ ($7.0 million, high)

A_3 = ($4.5 million, high) ~ ($4.5 million, high)

A_4 = ($5.0 million, average) ~ ($6.0 million, high)

Through the above tradeoff procedures, we can eliminate three attributes, X_3, X_5 and X_6. Then the original problem is left with three attributes with the following modified decision matrix:

$$D = \begin{bmatrix} 2. & \text{Mach} & 1500 \text{ NM} & \$4.5 \text{ million} \\ 1.875 & \text{Mach} & 2200 \text{ NM} & \$7.0 \text{ million} \\ 2.6 & \text{Mach} & 2200 \text{ NM} & \$4.5 \text{ million} \\ 2.2 & \text{Mach} & 1800 \text{ NM} & \$6.0 \text{ million} \end{bmatrix}$$

By simple observation A_3 dominates other alternatives and therefore is the preferred choice. If all alternatives are nondominated ones, we proceed to another higher order of tradeoffs until we are getting the single attribute values.

Note

A drawback of the hierarchical tradeoff analysis with two attributes at a time may be its slowness in reducing attributes. MacCrimmon and Wehrung [278] propose the lexicographic tradeoffs for eliminating this difficulty.

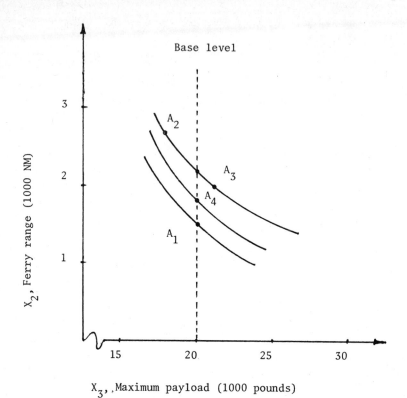

Fig. 3.11 A set of indifference curves between maximum payload and ferry range.

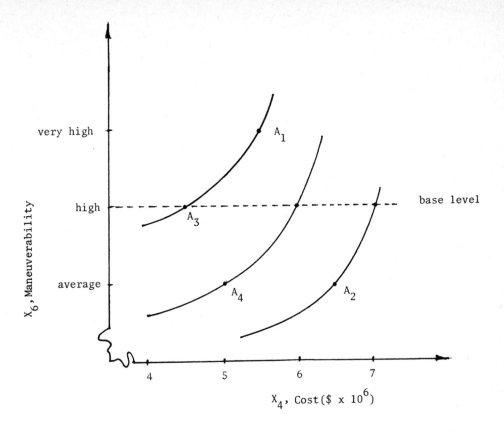

Fig. 3.12 A set of indifference curves between cost and maneuverability.

If the most important class by the lexicographic method (see Section 2.2.1) has more than one attribute, we form tradeoffs among these attributes. The second most important class of attributes is considered only if there are several alternatives having equally preferred attribute values in the most important class. This extended lexicography overcomes the noncompensatory characteristic of the standard lexicography by considering trade offs within a class.

Tradeoff information is more useful when designing multiple attribute alternatives than when choosing among final versions of them. MacCrimmon [274] gives a design example for transportation systems planning by the use of tradeoffs.

3. METHODS FOR INFORMATION ON ALTERNATIVE GIVEN

The methods in this class require that the DM be able to indicate his/her preference between two alternatives. This kind of information is far more demanding to assess than the information on attributes. LINMAP and nonmetric multidimensional scaling were originally developed to explain, rationalize, understand, and predict decision behavior, but they are well fitted for normative decision making.

3.1   METHODS FOR PAIRWISE PREFERENCE GIVEN

LINMAP and interactive simple additive weighting method take the pairwise preference information as an input. This input consists of a set of forced choices between pairs of alternatives. It is then expected that the set may contain inconsistent elements (choice). Both methods allow this inconsistency and try to minimize it.

actual alternatives?

3.1.1   LINMAP

Srinivasan and Shocker [380] developed LINMAP (LINear programming techniques for Multidimensional Analysis of Preference) for assessing the weights as well as locating the ideal solution. In this method m alternatives composed of n attributes are represented as m points in the n-dimensional space. A DM is assumed to have his ideal point that denotes the most preferred alternative location. Once the location of the ideal solution is decided, we can choose an alternative which has the shortest distance from the ideal solution. Given two alternatives, a DM is presumed to prefer an alternative which is closer to his ideal point. Then the weighted Euclidean distance, d_i, of the A_i from the ideal point is given by

$$d_i = \left[\sum_{j=1}^{n} w_j (x_{ij} - x_j^*)^2 \right]^{1/2}, \qquad i = 1, 2, \ldots, m \qquad (3.56)$$

where x_j^* is the ideal value for the jth attribute. The weights here take account of both the units in which each dimension is scaled and of the relative importance of each attribute to the DM. The square distance, $s_i = d_i^2$, is given by

$$s_i = \sum_{j=1}^{n} w_j (x_{ij} - x_j^*)^2, \qquad i = 1, 2, \ldots, m \qquad (3.57)$$

Let $\Omega = \{(k, \ell)\}$ denote a set of ordered pairs (k, ℓ) where k designates the preferred alternative on a forced choice basis resulting from a paired comparison involving k and ℓ. Usually Ω has $m(m - 1)/2$ elements in it. For every order pair $(k, \ell) \in \Omega$, the solution $(\underline{w}, \underline{x}^*)$ could be consistent with the weighted distance model if

$$s_\ell \geq s_k \tag{3.58}$$

Now the problem is to determine the solution $(\underline{w}, \underline{x}^*)$ for which the above conditions of (3.58) are violated as minimally as possible with the given decision matrix and Ω.

For the pair (k, ℓ), if $s_\ell \geq s_k$, the condition (3.58) is satisfied and there is no error attributable to the solution. On the other hand if $s_\ell < s_k$, then $(s_k - s_\ell)$ denotes the error extent to which the condition is violated. In general, if we define

$$(s_\ell - s_k)^- = \begin{cases} 0, & \text{if } s_\ell \geq s_k \\[2ex] s_k - s_\ell, & \text{if } s_\ell < s_k \end{cases} \tag{3.59}$$

$$= \max\ \{0,\ (s_k - s_\ell)\}$$

then $(s_k - s_\ell)^-$ denotes the error for the pair $(k, \ell) \in \Omega$. Summing it up for all the pairs in Ω, we obtain

B = poorness of fit

$$= \sum_{(k,\ell) \in \Omega} (s_\ell - s_k)^- \tag{3.60}$$

By definition, $(s - s_k)^-$ is nonnegative so that B is nonnegative. Our problem is to find a solution $(\underline{w}, \underline{x}^*)$ for which B is minimal. Usually this problem leads to the trivial solution with $w_j = 0$, hence we should add one constraint in minimizing B.

Let us define goodness of fit G, similar to B, such that

$$G = \sum_{(k,\ell)\in\Omega} (s_\ell - s_k)^+ \qquad (3.61)$$

where

$$(s_\ell - s_k)^+ = \begin{cases} s_\ell - s_k, & \text{if } s_\ell \geq s_k \\ 0, & \text{if } s_\ell < s_k \end{cases}$$

We then add the following condition, using the definition of G, in minimizing B,

$$G > B$$

or $\qquad (3.62)$

$$G - B = h$$

where h is an arbitrary positive number. By the definition of $(s_\ell - s_k)^-$ and $(s_\ell - s_k)^+$, it directly follows that

$$(s_\ell - s_k)^+ - (s_\ell - s_k)^- = s_\ell - s_k$$

Then eq. (3.62) can be written as

$$\sum_{(k,\ell)\in\Omega} (s_\ell - s_k) = h \qquad (3.63)$$

Thus $(\underline{w}, \underline{x}^*)$ can be obtained by solving the constrained optimization problem of

$$\min B = \sum_{(k,\ell)\in\Omega} \max\{0, (s_k - s_\ell)\}$$

s.t.

$$\sum_{(k,\ell)\in\Omega} (s_\ell - s_k) = h$$

This can be converted to the LP problem:

$$\text{min} \quad \sum_{(k,\ell)\in\Omega} z_{k\ell} \tag{3.64}$$

s.t.

$$(s_\ell - s_k) + z_{k\ell} \geq 0, \text{ for } (k, \ell) \in \Omega \tag{3.65}$$

$$\sum_{(k,\ell)\in\Omega} (s_\ell - s_k) = h \tag{3.66}$$

$$z_{k\ell} \geq 0, \qquad \text{for } (k, \ell) \in \Omega \tag{3.67}$$

The following reasoning may clear the doubt about the conversion. In LP formulation, (3.65) can be written as

$$z_{k\ell} \geq s_k - s_\ell$$

and due to its nonnegative

$$z_{k\ell} \geq \max\{0, (s_k - s_\ell)\} \tag{3.68}$$

Since we are minimizing the objective function, eq. (3.64), which is the sum of $z_{k\ell}$'s, eq. (3.68) will hold with equality at the optimum so that

$$z_{k\ell} = \max\{0, (s_k - s_\ell)\}$$

We know that $z_{k\ell}$'s are elements of B.

Finally we can make the LP formulation of eqs. (3.64) through (3.67) ready to solve by substituting eq. (3.57) into s_ℓ and s_k to obtain

$$s_\ell - s_k = \sum_{j=1}^{n} w_j (x_{\ell j} - x_j^*)^2 - \sum_{j=1}^{n} w_j (x_{kj} - x_j^*)^2$$

$$= \sum_{j=1}^{n} w_j (x_{\ell j}^2 - x_{kj}^2) - 2 \sum_{j=1}^{n} w_j x_j^* (x_{\ell j} - x_{kj}) \qquad (3.69)$$

Since x_j^* is an unknown constant, $w_j x_j^*$ is replaced by v_j. The proposed LP model is:

$$\min \sum_{(k,\ell) \in \Omega} z_{k\ell}$$

$$\text{s.t.} \sum_{j=1}^{n} w_j (x_{\ell j}^2 - x_{kj}^2) - 2 \sum_{j=1}^{n} v_j (x_{\ell j} - x_{kj}) + z_{k\ell} \geq 0, \text{ for } (k,\ell) \in \Omega$$

$$\sum_{j=1}^{n} w_j \sum_{(k,\ell) \in \Omega} (x_{\ell j}^2 - x_{kj}^2) - 2 \sum_{j=1}^{n} v_j \sum_{(k,\ell) \in \Omega} (x_{\ell j} - x_{kj}) = h \qquad (3.70)$$

$$w_j \geq 0, \ v_j \text{ unrestricted in sign}, \ j = 1, 2, \ldots, n$$

$$z_{k\ell} \geq 0, \text{ for } (k, \ell) \in \Omega$$

We may have these cases of LP solution:

i)   if $w_j^* > 0$, then $x_j^* = v_j^* / w_j^*$

ii) if $w_j^* = 0$ and $v_j^* = 0$, define $x_j^* = 0$

iii) if $w_j^* = 0$ and $v_j^* > 0$, then $x_j^* = +\infty$

iv) if $w_j^* = 0$ and $v_j^* < 0$, then $x_j^* = -\infty$

Then the square distance from the $\underline{x}^*$ is

$$s_i = \sum_{j'} w_{j'}^* (x_{ij'} - x_{j'}^*)^2 - 2 \sum_{j''} v_{j''}^* x_{ij''}, \quad i = 1,2,\ldots,m \qquad (3.71)$$

where

$$j' = \{j \mid w_j^* \geq 0\}$$

$$j'' = \{j \mid w_j^* = 0 \text{ and } v_j^* \neq 0\}$$

Srinivasan and Shocker [381] used another criteria for s_i in eq. (3.57). They used simple additive weighting model (2.2.1), then

$$s_i = - \sum_{j=1}^{n} w_j x_{ij}$$

where x_{ij}'s are at least intervally scaled and the larger x_{ij}, the larger preference. Then we have

$$s_\ell - s_k = \sum_{j=1}^{n} w_j (x_{kj} - x_{\ell j})$$

Now the corresponding model is:

$$\min \sum_{(k,\ell)\in\Omega} z_{k\ell}$$

$$\text{s.t.} \quad \sum_{j=1}^{n} w_j(x_{kj} - x_{\ell j}) + z_{k\ell} \geq 0, \text{ for } (k,\ell) \in \Omega$$

$$\sum_{j=1}^{n} w_j \sum_{(k,\ell)\in\Omega} (x_{kj} - x_{\ell j}) = h$$

$$w_j \geq 0, \qquad j = 1,2,\ldots,n \tag{3.72}$$

$$z_{k\ell} \geq 0, \text{ for } (k,\ell) \in \Omega$$

Numerical Example 1.

The following decision matrix with 2 attributes and 5 alternatives is considered [380].

$$D = \begin{array}{c} \\ A_1 \\ A_2 \\ A_3 \\ A_4 \\ A_5 \end{array} \begin{array}{cc} X_1 & X_2 \\ \left[\begin{array}{cc} 0 & 5 \\ 5 & 4 \\ 0 & 2 \\ 1 & 3 \\ 4 & 1 \end{array}\right] \end{array}$$

The forced-choice ordered paired comparison judgments are:

$$\Omega = \{(1, 2), (3, 1), (4, 1), (5, 1), (2, 3), (2, 4), (2, 5), (4, 3),$$
$$(3, 5), (4, 5)\}$$

That is, A_1 is preferred to A_2, A_3 is preferred to A_1, etc.

It is noted that there are intransitivities in the DM's preference judgment: for instance, $A_1 > A_2$ and $A_2 > A_3$, but $A_3 > A_1$, etc.

To obtain the best weight and ideal point, a linear programming problem is set up using eq. (3.70). The first of the constraints of eq. (3.70) is obtained from the paired comparison $(k, \ell) = (1, 2)$:

$$\sum_{j=1}^{2} w_j (x_{2j}^2 - x_{1j}^2)^2 - 2 \sum_{j=1}^{2} v_j (x_{2j} - x_{1j}) + z_{12} \geq 0$$

i.e.,

$$w_1 (5^2 - 0)^2 + w_2 (4^2 - 5^2) - 2v_1 (5 - 0) - 2v_2 (4 - 5) + z_{12} \geq 0$$

$$25w_1 - 9w_2 - 10v_1 + 2v_2 + z_{12} \geq 0$$

By similar calculations, the linear programming problem of eq. (3.70) becomes:

$$\min \quad z = z_{12} + z_{31} + z_{41} + z_{51} + z_{23} + z_{24} + z_{25} + z_{43} + z_{35} + z_{45}$$

s.t.
$$25w_1 - 9w_2 - 10v_1 + 2v_2 + z_{12} \geq 0$$
$$21w_2 - 6v_2 + z_{31} \geq 0$$
$$-w_1 + 16w_2 + 2v_1 - 4v_2 + z_{41} \geq 0$$
$$-16w_1 + 24w_2 + 8v_1 - 8v_2 + z_{51} \geq 0$$
$$-25w_1 - 12w_2 + 10v_1 + 4v_2 + z_{23} \geq 0$$
$$-24w_1 - 7w_2 + 8v_1 + 2v_2 + z_{24} \geq 0$$
$$-9w_1 - 15w_2 + 2v_1 + 6v_2 + z_{25} \geq 0$$
$$-w_1 - 5w_2 + 2v_1 + 2v_2 + z_{43} \geq 0$$
$$16w_1 - 3w_2 - 8v_1 + 2v_2 + z_{35} \geq 0$$
$$15w_1 - 8w_2 - 6v_1 + 4v_2 + z_{45} \geq 0$$
$$-20w_1 + 2w_2 + 8v_1 + 4v_2 = 1$$

$$w_1, w_2, z_{12}, z_{31}, z_{41}, z_{51}, z_{23}, z_{24}, z_{25}, z_{43}, z_{35}, z_{45} \geq 0$$

$$v_1, v_2 \text{ unrestricted in sign}$$

The optimal solution is found to be

$z^* = .25$

$z_{12} = .25$, $z_{31} = z_{41} = z_{51} = z_{23} = z_{24} = z_{25} = z_{43} = z_{35} = z_{45} = 0$

$\underline{w}^* = (.0277, .0554)$

$\underline{v}^* = (.0833, .1944)$

The weights can be multiplied by an arbitrary positive scalar. Choosing this scale to be $1/.0277$, we obtain

$\underline{w}^* = (1., 2.)$

$\underline{v}^* = (3., 7.)$

$\underline{x}^* = (v_1^*/w_1^*, v_2^*/w_2^*) = (3., 3.5)$

The square distance from $\underline{x}^*$ using eq. (3.71) is

$\underline{s} = (13.5, 4.5, 13.5, 4.5, 13.5)$

The above distance s_i's satisfy all the partial orders in Ω except $(1, 2)$. The fact that $z_{12} = .25$ indicates that the weights imply $s_1 > s_2$, and hence the judgment that A_1 is preferred to A_2 is violated by the optimal solution. The DM may select either A_2 or A_4 which has the minimum distance from the ideal point.

Numerical Example 2 (The Fighter Aircraft Problem)

The decision matrix after assigning numerical values (by using the interval scale of Fig. 2.4) to qualitative attributes is

$$
D = \begin{array}{c} \\ A_1 \\ A_2 \\ A_3 \\ A_4 \end{array}
\begin{array}{cccccc}
X_1 & X_2 & X_3 & X_4 & X_5 & X_6 \\
\left[\begin{array}{cccccc}
2.0 & 1500 & 20000 & 5.5 & 5 & 9 \\
2.5 & 2700 & 18000 & 6.5 & 3 & 5 \\
1.8 & 2000 & 21000 & 4.5 & 7 & 7 \\
2.2 & 1800 & 20000 & 5.0 & 5 & 5
\end{array}\right]
\end{array}
$$

Since a set of weights in LINMAP takes account of the units in which each attribute is scaled as well as the relative importance of each attribute, the decision matrix can be rewritten as

$$
D' = \left[\begin{array}{cccccc}
2.0 & 1.5 & 2.0 & 5.5 & 5 & 9 \\
2.5 & 2.7 & 1.8 & 6.5 & 3 & 5 \\
1.8 & 2.0 & 2.1 & 4.5 & 7 & 7 \\
2.2 & 1.8 & 2.0 & 5.0 & 5 & 5
\end{array}\right]
$$

where X_2 is expressed by 10^3 NM and X_3 by 10^4 pounds.

The ordered paired comparison judgments from the DM are:

$$
\Omega = \{(1, 2), (1, 3), (4, 1), (3, 2), (2, 4), (3, 4)\}
$$

Then the LP formulation of LINMAP using eq. (3.70) becomes

$$\min z = z_{12} + z_{13} + z_{41} + z_{32} + z_{24} + z_{34}$$

s.t.

$$
\begin{bmatrix} w_1 \\ w_2 \\ w_3 \\ w_4 \\ w_5 \\ w_6 \\ v_1 \\ v_2 \\ v_3 \\ v_4 \\ v_5 \\ v_6 \\ z_{12} \\ z_{13} \\ z_{41} \\ z_{32} \\ z_{24} \\ z_{34} \end{bmatrix}^T
\begin{bmatrix}
2.25 & -.76 & -.84 & 3.01 & -1.41 & 1.6 & 3.85 \\
5.04 & 1.75 & -.99 & 3.29 & -4.05 & -.76 & 4.28 \\
-.76 & .41 & 0. & -1.17 & .76 & -.41 & -1.17 \\
12. & -10. & 5.25 & 22. & -17.25 & 4.75 & 16.75 \\
-16. & 24. & 0. & -40. & 16. & -24. & -40. \\
-56. & -32. & 56. & -24. & 0. & -24. & -80. \\
-1. & .4 & .4 & -1.4 & .6 & -.8 & -1.8 \\
-2.4 & -1. & .6 & -1.4 & 1.8 & .4 & -2. \\
.4 & -.2 & 0. & .6 & -.4 & .2 & .6 \\
-2. & 2. & -1. & -4. & 3. & -1. & -3. \\
4. & -4. & 0. & 8. & -4. & 4. & 8. \\
8. & 18. & -8. & 4. & 0. & 4. & 26. \\
1 & & & & & & \\
& 1 & & & & & \\
& & 1 & & & & \\
& & & 1 & & & \\
& & & & 1 & & \\
& & & & & 1 & \\
\end{bmatrix}
\geq
\begin{bmatrix} 0 \\ 0 \\ 0 \\ 0 \\ 0 \\ 0 \\ 1 \end{bmatrix}
$$

$$w_j \geq 0, \qquad j = 1,2,\ldots,6$$

v_j is unrestricted in sign, $\qquad j = 1,2,\ldots,6$

$$z_{12}, z_{13}, z_{14}, z_{32}, z_{24}, z_{34} \geq 0$$

The optimal solution is

$$\underline{z}^* = .5 \qquad (z_{41}^* = .5, z_{12}^* = z_{13}^* = z_{32}^* = z_{24}^* = z_{34}^* = 0.)$$

$$\underline{w}^* = (0., 0., 0., 0., 0., 0.0625)$$

$$\underline{v}^* = (0., 0., 0., 0., 0., .5)$$

and the ideal point is obtained as

$$\underline{x}^* = (0., 0., 0., 0., 0., 8.)$$

The square distances from the $\underline{x}^*$ are

$$s_i = w_6^*(x_{i6} - x_6^*)^2, \qquad i = 1,2,3,4$$

which is

$$\underline{s} = (.0625, .5625, .0625, .5625)$$

The DM may select A_1 or A_3 which has the minimum distance from the ideal point.

This problem is tried again by the simple additive weighting model of LINMAP for comparison purposes. The cost attribute, X_4, is changed to the benefit one by taking the negative value of it. The changed decision matrix is now

$$D'' = \begin{bmatrix} 2.0 & 1.5 & 2.0 & -5.5 & 5 & 9 \\ 2.5 & 2.7 & 1.8 & -6.5 & 3 & 5 \\ 1.8 & 2.0 & 2.1 & -4.5 & 7 & 7 \\ 2.2 & 1.8 & 2.0 & -5.0 & 5 & 5 \end{bmatrix}$$

The LP formulation of LINMAP using eq. (3.72) is

$$\min z = z_{12} + z_{13} + z_{41} + z_{32} + z_{24} + z_{34}$$

s.t.
$$
\begin{array}{llll}
-.5w_1 - 1.2w_2 + .2w_3 + w_4 + 2w_5 + 4w_6 + z_{12} & & & \geq 0 \\
.2w_1 - .5w_2 - .1w_3 - w_4 - 2w_5 + 2w_6 & + z_{13} & & \geq 0 \\
.2w_1 + .3w_2 + .5w_4 - 4w_6 & & + z_{41} & \geq 0 \\
-.7w_1 - .7w_2 + .3w_3 + 2w_4 + 4w_5 + 2w_6 & & + z_{32} & \geq 0 \\
.3w_1 + .9w_2 - .2w_3 - 1.5w_4 - 2w_5 & & + z_{24} & \geq 0 \\
-.4w_1 + .2w_2 + .1w_3 + .5w_4 + 2w_5 + 2w_6 & & + z_{34} & \geq 0 \\
-.9w_1 - 1.0w_2 + .3w_3 + 1.5w_4 + 4w_5 + 6w_6 & & & = 1
\end{array}
$$

$$w_j \geq 0, \qquad j = 1,2,\ldots,6$$
$$z_{12}, z_{13}, z_{41}, z_{32}, z_{24}, z_{34} \geq 0$$

The optimal solution is

$$\underline{z}^* = .5 \ (z_{41} = 0.5, \ z_{12} = z_{13} = z_{32} = z_{24} = z_{34} = 0.)$$

$$\underline{w}^* = (0., \ .2, \ 0., \ 0., \ .09, \ .14)$$

The values of s_i using

$$s_i = - \sum_{j=1}^{6} w_j x_{ij}$$

are

$$\underline{s} = (-2.01, \ -1.51, \ -2.01, \ -1.51)$$

Again alternative A_1 and A_3 have the minimum value of s_i which matches the result obtained by using eq. (3.70).

Note

The LINMAP procedure does not require that the set Ω consist of all $m(m-1)/2$ paired comparison judgments from the DM. However, the set of weights obtained by LINMAP will be more reliable if the number of pairs in Ω is large. When the number of alternatives is greater than the number of attributes (i.e., $m > n$), LINMAP gives the better fitting. The method does not require that the paired comparison judgment be transitive.

It is noted that the LINMAP procedure using eq. (3.70), which is a weighted Euclidean distance from the ideal point, is based on the quadratic utility function [469]. For an example of quadratic utility, consider the utility (or value) of the number of rooms when one is purchasing a house. The buyer likes neither the small number of rooms nor the large number. He may prefer 4 to 5 or 3. As in this case, the best value of some attributes may be located in the middle of the attribute range.

LINAMP also takes into account a monotonically increasing (or decreasing) utility function where the ideal point is located at the positive or negative infinite. Recall that $v_j^* = w_j^* x_j^*$ and $x_j^* = v_j^*/w_j^*$ in eq. (3.71), then the set j" is identified to be the set of attributes whose ideal values are located at positive or negative infinite. Hence if the DM uses a monotonically increasing (or decreasing) utility function only, the set j" will represent the set of whole attributes {j}. The value, s_i, then becomes

$$s_i = -2 \sum_{j=1}^{n} v_j x_{ij}, \qquad i = 1,2,\ldots,m$$

where v_j is interpreted as the "salience factor" associated with X_j. This situation is equivalent to using eq. (3.72), the simple additive weighting model. When all attributes in the decision matrix are recognized to take the monotonic utility, it is simpler computationally to use LINMAP of eq. (3.72). The same solutions for the fighter aircraft problem obtained by using eq. (3.70) and eq. (3.72) indicate that the DM made paired comparison judgments based on the linear rather than the quadratic utility.

In ELECTRE (2.3.5) the necessary input is a set of weights, and the output is a set of outranking relationships (or partial orders); whereas for LINMAP the DM gives a set of the partial orders as input, and a set of weights is the output. When the number of attributes exceeds the number of alternatives (i.e., n > m), it is not easy for the DM to assess the partial orders, and it is hard to obtain reliable weights by the LINMAP; therefore, in this case it is better to use ELECTRE.

3.1.2 INTERACTIVE SIMPLE ADDITIVE WEIGHTING METHOD

Kornbluth [249] presents an interactive method for ranking alternatives subject to an (initially) unspecified linear utility function. The
method is economical in the number of paired judgments that must be made
by the DM and leads to the identification of the final ranking and the
space of weights for the corresponding linear utility functions which
would lead to this ranking.

Assume that the linear utility function (i.e., simple additive
weighting) be adopted for the decision analysis, then an alternative p
is preferred to q if,

$$\sum_j w_j x_{pj} > \sum_j w_j x_{qj} \tag{3.73}$$

In the vector form eq. (3.73) becomes

$$\underline{w}^T (\underline{x}_p - \underline{x}_q) > 0 \tag{3.74}$$

where $\underline{w} \in W = \{\underline{w} \mid \sum_{i=1}^n w_i = 1, w_i > 0\}$.

Let Ω be a permutation of m alternatives, which represents a preference order
of the m alternatives, and $\Omega(j)$ be the alternative in the jth position of the
order Ω. Then eq. (3.74) induces the following (m - 1) inequality relations,

$$\underline{w}^T (\underline{x}_{\Omega(j)} - \underline{x}_{\Omega(j+1)}) > 0, \qquad j = 1, 2, \ldots, m - 1 \tag{3.75}$$

Let $W_\Omega = \{\underline{w} \mid \underline{w} \in W, \underline{w}$ satisfies eq. (3.75)}. If Ω is the ordering preferred
by the DM, he can be indifferent to the use of any $\underline{w} \in W_\Omega$ as the weights for
the linear utility function since any such w will produce the order Ω [248].
Conversely if $\underline{w} \in W_\Omega$ is acceptable by the DM as the appropriate set of weights

for such a utility function he should accept the order Ω as his preference
ordering. The interactive simple additive weighting method introduces a
way whereby the DM can progressively change Ω and approach his desired
order Ω^* and the associated $\underline{w}$ space $W_\Omega \ \varepsilon \ W$.

We note that the set of all the feasible orderings $\{\Omega\}$ induces a parti-
tion of W, i.e.,

$$W = \underset{\Omega}{U} \ W_\Omega$$

and

$$W_\Omega \ \cap \ W_{\Omega'} \ = \ \emptyset, \qquad \Omega \neq \Omega'$$

and that in defining W_Ω it is only necessary to consider constraints formed
by adjacent pairs of elements in Ω. The reminder are trivially satisfied.
Hence each set W_Ω is determined by the set of linear constraints (3.75),
and in particular by the tight constraints for which equality can hold in
eq. (3.75). An order Ω will be called   the infeasible order if any of
the following constraints are satisfied:

$$\underline{x}_{\Omega(j)} - \underline{x}_{\Omega(j+1)} < \underline{0}, \quad j = 1, 2, \ldots, m-1 \tag{3.76}$$

In other words, if any one of the constraint set of eq. (3.76) is satisfied,
no $\underline{w}$ can produce the order Ω. Simply the dominated alternative cannot be
ranked higher than the dominant one. Only for the nondominated pair, alternatives can be ranked alternately due to the different $\underline{w}$.

Assume that the A_k and A_ℓ are adjacent in the current order Ω, and that
the constraint:

$$\underline{w}^T (\underline{x}_k - \underline{x}_\ell) > 0 \tag{3.77}$$

$$\underline{w} \ \varepsilon \ W_\Omega$$

is a tight constraint.

The operation of switching the order of k,ℓ in Ω to ℓ,k in Ω' is equivalent to moving from the space W_Ω to an adjacent space $W_{\Omega'}$, across the boundary determined by eq. (3.77). If Ω is a feasible order then Ω' will also be feasible. This finding allows the DM to correct and improve the order Ω by switching a pair of the order at a time. If he corrects only the binding constraints (order), the method maintains feasibility through the process; further, the amount of material that the DM need review is kept to a minimum.

Now the problem is to identify the tight constraint set from $(m-1)$ set of eq. (3.75). Given the feasible order Ω, let B_Ω be the $(m-1)$ x m matrix whose rows $B_{\Omega(i)}$ are given by $x_{\Omega(i)} - x_{\Omega(i+1)}$. The closure of W_Ω is the set $\{\underline{w}\}$ such that

$$B_\Omega \underline{w} \geq 0 \qquad (3.78)$$

$$\sum_{i=1}^{n} w_i = 1 \qquad (3.79)$$

$$w_i \geq 0, \quad i = 1, 2, \ldots, n \qquad (3.80)$$

The boundary of W_Ω is determined by those rows of eq. (3.78) which are tight (active) constraints for some value of $\underline{w}$. For such rows the optimum value of the following LP problem gives zero:

$$\begin{aligned}
&\min \ B_{\Omega(i)} \ \underline{w} \\
\text{s.t.} \\
&B_\Omega \ \underline{w} \geq 0 \\
&\sum_{i=1}^{n} w_i = 1 \\
&w_i \geq 0, \quad i = 1, 2, \ldots, n
\end{aligned} \qquad (3.81)$$

Since we expect the number of alternatives m will, in general, be greater than the number of attributes n, it is more convenient to consider the dual problem of the LP formulation of eq. (3.81).

$$
\begin{aligned}
&\max \rho \\
&\text{s.t. } B_\Omega^T \underline{\mu} + \rho \underline{1} \le B_{\Omega(i)}^T \\
&\quad \underline{\mu} > \underline{0}, \quad \rho \text{ unconstrained} \\
&\text{where} \quad \underline{1} \text{ is } (n \times 1) \text{ unit column vector} \\
&\quad \underline{\mu}^T = (\mu_1, \mu_2, \ldots, \mu_{m-1})
\end{aligned} \tag{3.82}
$$

By LP duality, if eq. (3.82) has a feasible solution for which $\rho^* = 0$, then the row $B_{\Omega(i)}$ represents a binding constraint for W_Ω and the associated pair $(\Omega_{(i)}, \Omega_{(i+1)})$ represents a binding pair. Furthermore, a strictly positive value for any variable μ_k at an optimum of LP formulation of eq. (3.82) implies that $B_{\Omega(k)}$ is a binding constraint. This fact can be used to reduce the number of dual problems that need to be solved in order to determine the set of binding pairs of elements in Ω.

The method suggested for analyzing the DM's preferences and determining Ω^* is as follows:

Step 1. Identify an initial feasible ordering (say lexicographical ordering or equal weights on all criteria).

Step 2. Identify the binding constraints of W_Ω.

Step 3. Present the DM with the list of pairs in Ω which determine the boundary of W_Ω (and their associated characteristics).

Stop if this order is accepted by the DM, otherwise ask the DM to switch one pair from the list and return to step 2.

Numerical Example

Consider the following decision matrix D of 7 alternatives with 3 attri-butes. All attributes are the benefit criteria (the greater, the higher preference), and they are transformed into the common scale.

$$
D = \begin{array}{c}
 \begin{array}{ccc} X_1 & X_2 & X_3 \end{array} \\
\left[\begin{array}{ccc}
5 & 8 & 0 \\
4 & 0 & 6 \\
2 & 4 & 3 \\
8 & 1 & 4 \\
2 & 7 & 1 \\
6 & 4 & 2 \\
3 & 2 & 6
\end{array}\right]
\begin{array}{c}
A_1 \\ A_2 \\ A_3 \\ A_4 \\ A_5 \\ A_6 \\ A_7
\end{array}
\end{array}
$$

Iteration No. 1:

Step 1. Identify the initial ordering: We use lexicographic method (see Section 2.2.1): X_1 is the most important, X_2 is next, then $\Omega_1 = (A_4, A_6, A_1, A_2, A_7, A_5, A_3)$ is a feasible order.

Step 2. Identify the binding constraints of W_1: The associated B_1 matrix is:

$$
B_1 = \begin{bmatrix}
B_{11} \\ B_{12} \\ B_{13} \\ B_{14} \\ B_{15} \\ B_{16}
\end{bmatrix}
\begin{bmatrix}
X_4 - X_6 \\ X_6 - X_1 \\ X_1 - X_2 \\ X_2 - X_7 \\ X_7 - X_5 \\ X_5 - X_3
\end{bmatrix}
=
\begin{bmatrix}
2 & -3 & 2 \\
1 & -4 & 2 \\
1 & 8 & -6 \\
1 & -2 & 0 \\
1 & -5 & 5 \\
0 & 3 & -2
\end{bmatrix}
$$

In order to identify the space W_1 we solve a series of LP problems:

max ρ

s.t.

$$B_1^T \begin{bmatrix} \mu_1 \\ \mu_2 \\ \cdot \\ \cdot \\ \cdot \\ \mu_6 \end{bmatrix} + \rho \begin{bmatrix} 1 \\ 1 \\ 1 \end{bmatrix} \leq B_{1i}^T$$

$$\mu_i \geq 0, \quad i = 1, 2, \ldots, 6$$

ρ unconstrained.

where B_{1i}^T, $i = 1, 2, \ldots, 6$ are rows of B_i.

For B_{11} the solution is $\rho^* = .75$. $\mu_2 = .625$ and $\mu_4 = .625$ are basic (strictly positive) implying that the pairs (6,1) (2,7) are binding. The pair (4,6) is not binding. Since (6,1) is already identified as a binding pair we do not need to solve for B_{12}. For B_{13} the solution is $\rho^* = .2222$. $\mu_4 = .7778$ and $\mu_6 = 3.111$ are basic implying that the pairs (2,7) (5,3) are binding, but (1,2) is not, etc. The binding constraints are: (6,1), (2,7), (7,5), (5,3).

Step 3. The decision step: The DM is unsatisfied with the relative positions of alternatives 6 and 1, and suggests that A_1 should be ranked above A_6. Now $\Omega_2 = (4, 1, 6, 2, 7, 5, 3)$. Return to Step 2.

Iteration No. 2:

Step 2. Identify the binding constraint of W_2: The resulting B_2 matrix is

$$B_2 = \begin{bmatrix} B_{21} \\ B_{22} \\ B_{23} \\ B_{24} \\ B_{25} \\ B_{26} \end{bmatrix} = \begin{bmatrix} x_4 - x_1 \\ x_1 - x_6 \\ x_6 - x_2 \\ x_2 - x_7 \\ x_7 - x_5 \\ x_5 - x_3 \end{bmatrix} = \begin{bmatrix} 3 & -7 & 4 \\ -1 & 4 & -2 \\ 2 & 4 & -4 \\ 1 & -2 & 0 \\ 1 & -5 & 5 \\ 0 & 3 & -2 \end{bmatrix}$$

Also the LP formulation is

$$\max \rho$$

s.t.

$$B_2^T \begin{bmatrix} \mu_1 \\ \mu_2 \\ \vdots \\ \mu_6 \end{bmatrix} + \rho \begin{bmatrix} 1 \\ 1 \\ 1 \\ \end{bmatrix} \leq B_{2i}^T$$

$$\mu_i > 0, \quad i = 1, 2, \ldots, 6$$

ρ unconstrained

First we know pair $(1,6)$ is binding from the judgment in Iteration No. 1. For B_{21} the solution is $\rho^* = .3889$. $\mu_4 = 1.889$ and $\mu_5 = .7222$ are basic implying $(4,1)$ is not binding but $(2,7)$ and $(7,5)$ are binding. For B_{26} the solution is $\rho^* = .25$. $\mu_2 = 1.125$ and $\mu_4 = .875$ implying $(2,7)$ is binding but not $(5,3)$. The only undecided pair is $(6,2)$. For B_{23} the solution is $\rho^* = 1.0$. The pair $(6,2)$ is not binding. The binding constraints are: $(1,6)$, $(2,7)$, $(7,5)$.

Step 3. The decision step: The DM accepts the present order $\Omega_2 = (4, 1, 6, 2, 7, 5, 3)$. The associated weight is decided by the above three binding pairs;

$$\begin{bmatrix} -1 & 4 & -2 \\ 1 & -2 & 0 \\ 1 & -5 & 5 \end{bmatrix} \begin{bmatrix} w_2 \\ w_4 \\ w_5 \end{bmatrix} \geq \underline{0}$$

$$w_2 + w_4 + w_5 = 1$$

$$w_i \geq 0, \quad i = 2,4,5$$

A solution of the above constraint is $\underline{w} = (.6, .25, .15)$. Conversely we may verify that this $\underline{w}$ can reproduce Ω_2 by the simple additive weighting method.

Note

In the simulation tests on random data which are carried out by Kornbluth [249], it is shown that the number of comparisons that must be made at each stage tends to be less than n+1, where n is the number of attributes being considered. Simulation results also suggest that in cases where the number of attributes is 6 or less, a maximum of 20 to 30 alternatives need be considered in order to obtain a reasonable estimate of the appropriate weight space.

3.2   METHOD FOR PAIRWISE PROXIMITY GIVEN

The DM's ordering of the proximities of pairs of alternatives can be used to construct a multidimensional spatial representation. Alternatives are represented by points in the space of much less dimensions than the original. The DM is asked to locate ideal alternatives in this space, and then the distance from the ideal point is measured.

3.2.1   MULTIDIMENSIONAL SCALING WITH IDEAL POINT

Multidimensional scaling (MDS) has been developed mainly in the psycho-metric literatures [252, 253, 366] in order to find the 'hidden structure' in the experimental data. Recently MDS technique has been widely utilized in   marketing problems, e.g., for predicting consumers' preferences [BM-8, BM-9, BM-10]. When alternatives have too many attributes or in some cases have vague or unknown attributes, MDS has great advantage in the solution of MADM problems.

In MDS the DM's orderings of the proximities of pairs of alternatives can be used to construct a multidimensional spatial representation. Al-ternatives are represented by points in the space. The points that are close together are assumed to be close together in terms of preference. The DM is asked to locate his ideal point in the space and then the distance from the ideal point is measured (using a Euclidean or other measure) in order to rank the alternatives in terms of preference. Some key features of MDS will be described because of its descriptive emphasis and its complexity of analysis. Kruskal and Green's text deal with this topic in detail [BM-10, BM-14].

Nonmetric MDS: Suppose you are given a map showing the locations of several cities in the United States, and are asked to construct a table of distances between these cities. It is a simple matter to fill in any entry in the table by measuring the distance between the cities with a ruler, and converting the ruler distance into the real distance by using the scale of the map. Now consider the reverse problem, where you are given the table of distances between the cities which are identified by numbers, and are asked to produce the map. MDS is a method for solving this reverse problem. This ex-ample is only confined to   two dimensional scaling, but MDS can offer any dimensional scaling as the name indicates. The procedure for converting from the ratio scaled midpoint distances to the mapping of the cities is called metric MDS.

Because both input and output are composed of real ratio values (i.e. their pairwise relationships can be expressed as a regular mathematical function), Shepard [366] suggested replacing the ratio scaled proximities (or distances) with rank order dissimilarities as an input. For this reason, the name nonmetric MDS is used to describe MDS which is designed to find a configuration whose order of (ratio-scaled) interpoint distances best produces the original rank order of the input dissimilarities. Kruskal [252, 253, 255] developed this idea and suggested a practical algorithm to arrive at a final configuration.

The nonmetric MDS is a procedure of representing m alternatives geometrically by m points in space, so that interpoint distances correspond to rank order of dissimilarity judgments between alternatives. Hence the problem is how to find the best-fitting configuration in chosen t-dimensional space. Let us denote the DM's dissimilarity judgment between alternatives i and j by δ_{ij}. We suppose that the judgment is inherently symmetrical, so that $\delta_{ij} = \delta_{ji}$. We also ignore the self-dissimilarities δ_{ii}. Thus with m alternatives, there are $m(m-1)/2$ numbers, namely δ_{ij} for $i < j$, $i = 1, 2, \ldots, m-1$, $j = 2, 3, \ldots, m$. We ignore the possibility of ties, that is, we assume that no two of these $m(m-1)/2$ numbers are equal. Since we assume no ties, it is possible to rank the dissimilarities in strictly ascending order,

$$\delta_{i_1 j_1} < \delta_{i_2 j_2} < \ldots < \delta_{i_M j_M}$$

where $M = m(m-1)/2$. We now wish to represent the m alternatives by m points in t-dimensional space. Let us call these points $x_1, x_2, \ldots, x_m$, and let d_{ij} denote the distance from x_i to x_j; then we have

$$d_{ij} = \sqrt{\sum_{s=1}^{t} (x_{is} - x_{js})^2}$$

In order to see how well the distances d_{ij} match the dissimilarities δ_{ij}, let us first ask 'What should perfect match mean?' Surely it should mean that whenever one dissimilarity is smaller than another, then the corresponding distances satisfy the same relationship. In other words, perfect match should mean that if we lay out the distances d_{ij} in an array

$$d_{i_1 j_1}, \; d_{i_2 j_2}, \; d_{i_3 j_3}, \; \ldots, \; d_{i_M j_M}$$

corresponding to the array of dissimilarity given above, then the smallest distance comes first, and the other distances follow in ascending order. Let us consider an example of $m = 4.$ There are $M = 6$ stars in the scatter diagram (Fig. 3.13). Each star corresponds to a pair of points as shown. Star (i, j) has abscissa d_{ij} and ordinate δ_{ij}. In terms of the scatter diagram, the perfect match means that as we trace out the stars one by one from bottom to top, we always move to the right, never to the left. This fails in Fig. 3.13 but holds in Fig. 3.14. To measure how far a scatter diagram such as Fig. 3.13 departs from the ideal of perfect fit, it is natural to fit an ascending curve to the stars as in Fig. 3.13, and then to measure the deviation from the star to the curve. Kruskal [252] suggested 'stress' formulas for the index of fitness as

$$S = \left[\frac{\sum\limits_{i \neq j}^{m} (d_{ij} - \hat{d}_{ij})^2}{\sum\limits_{i \neq j}^{m} d_{ij}^2} \right]^{1/2}$$

where $\hat{d}_{ij}$ is a set of ratio-scaled numbers, chosen to be as close to their

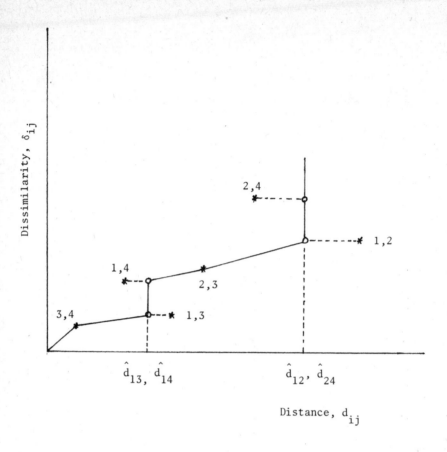

Fig. 3.13   An ascending curve on the initial configuration showing
the value of $\hat{d}_{ij}$.

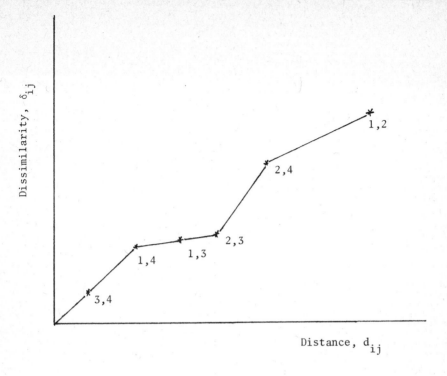

Fig. 3.14   A perfect configuration.

respective d_{ij} as possible, subject to being monotonical with δ_{ij}. It is known

that the chances of finding proper fitting increases (i.e., stress decreases)

as the configurational dimension increases, but it becomes difficult to

interpret   the hidden structure from that configuration. Hence the pro-

blem is to find the lowest dimensionality for which the monotonicity con-

straint is satisfied as far as possible. Kruskal also suggested the following

verbal evaluation for stress determination:

Stress	Goodness of fit
20%	poor
10%	fair
5%	good
2.5%	excellent
0%	perfect

Linear Regression for Dimensional Interpretation [BM-14]: Once the

appropriate configuration is made, the next problem is how to interpret the

relative positions in the space. In other words, positions in the confi-

guration may be systematically associated with some characteristics of

the attributes which were scaled. One way to discover these characteri-

stics is simply by looking at the configuration and recalling what is

known about the alternatives. But this is difficult due to its ambiguity.

The method which is the easiest to understand and most

commonly used is based on linear regression. Suppose we have some variables

associated with the attributes which, we suspect, may have a systematic

relationship to position in the configuration. One way to see if it does

is to perform a linear multiple regression using this variable as the

dependent variable and the coordinates of the configuration as the inde-

pendent variables. In other words we are regressing the attribute over

the coordinates of the configuration. What this means is that we seek

some weighted combination of the coordinates of the configuration which agrees with or explains the attribute as well as possible. The multiple

correlation coefficient is one measure of how well this can be done.

Identification of Ideal Point: Once the proper interpretation of

dimensions are made, the DM is supposed to locate his ideal point on the

chosen configuration. When the dimension of the configuration is greater

than two, it is hard to expect the DM to identify the ideal point.

This identification task can be made easier by following the LINMAP

procedures (see section 3.1.1); then the nonmetric MDS is used simply for

the reduction of the original dimensionality.

Numerical Example 1 [147]

To illustrate the MDS concepts, assume that a DM has been given the task

of judging pairs of cars   (models as shown in Fig.3.15) in terms of over-

all similarity. To be specific, suppose that the DM is shown a set of $(_{11}C_2)$

cards on each of which are the names of two cars. Using criteria of his own

choosing--but maintaining these criteria over all trials--he first places

the cards into two piles, representing similar cars versus different cars.

This process may be repeated, leading to a set of classes ordered by decreasing

similarity. Finally the DM is asked to rank order the cards within each

subpile in terms of decreasing similarity. The ultimate result is a rank

order of all 55 pairs, as in Table 3.4.

Suppose we try to represent the entries for which ordered dissimilarities

are shown in Table 3.4 as a geometric configuration whose rank order of inter-

point distances (in the configuration) display, for a specified number of

dimensions, close correspondence to the rank order of the dissimilarities

shown in Table 3.4. Such a configuration appears in Fig. 3.15. Note that the

interpoint distance between Continental (pt. 3) and Imperial (pt.6) is the

shortest, but that between Falcon (pt. 5) and Continental (pt. 3) is the longest.

What has been accomplished is the con-

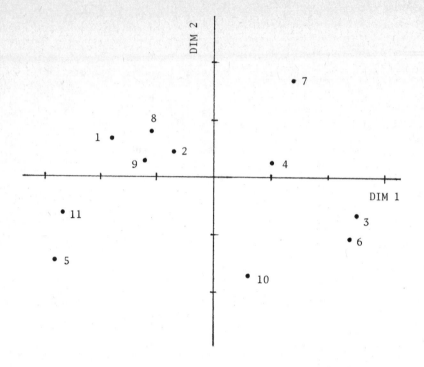

Alternatives--1968 Car Models

1.   Ford Mustang 6
2.   Mercury Cougar V8
3.   Lincoln Continental V8
4.   Ford Thunderbird V8
5.   Ford Falcon 6
6.   Chrysler Imperial V8
7.   Jaguar Sedan
8.   AMC Javelin V8
9.   Plymouth Barracuda V8
10.   Buick Le Sabre V8
11.   Chevrolet Corvair

Fig. 3.15   Two dimensional configuration for 11 car models [147]

Table 3.4   Rank order of dissimilarities between pairs of car models* [147]

A_i	1	2	3	4	5	6	7	8	9	10	11
1	-										
2	8	-									
3	50	38	-								
4	31	9	11	-							
5	12	33	55	44	-						
6	48	37	1	13	54	-					
7	36	22	23	16	53	26	-				
8	2	6	46	19	30	47	29	-			
9	5	4	41	25	28	40	35	3	-		
10	39	14	17	18	45	24	34	27	20	-	
11	10	32	52	42	7	51	49	15	21	43	-

*The rank number '1' represents the most similar pair.

version of rank order data on pairs of points to ratio-scaled data
on interpoint distances, while the axes of the configuration can be viewed
as intervals scaled with common units.

Numerical Example 2

The Queen Motor Company wants to select a sales representative for dis-
patching to the European market. The sales manager recommended 7 candidates
for the opening and evaluated them on 6 attributes in the following decision
matrix:

$$
D = \begin{bmatrix}
8 & 4 & 7 & 5 & 0 & 4 \\
0 & 1 & 4 & 4 & 6 & 1 \\
4 & 6 & 3 & 2 & 3 & 7 \\
1 & 3 & 2 & 8 & 4 & 5 \\
7 & 2 & 8 & 2 & 1 & 2 \\
4 & 9 & 5 & 6 & 2 & 3 \\
2 & 5 & 1 & 3 & 6 & 8
\end{bmatrix}
\begin{matrix}
A_1 \\ A_2 \\ A_3 \\ A_4 \\ A_5 \\ A_6 \\ A_7
\end{matrix}
$$

with columns X_1, X_2, X_3, X_4, X_5, X_6.

where the outcomes are based on the 10-point bipolar scale with the-larger-
the-greater preferences. The president of the company cannot depend solely
on the decision matrix arranged by his sales manager, so he himself inter-
viewed 7 candidates. The president said "I cannot easily judge which one is
the best, but I may make the dissimilarity judgment between the pairs of
candidates" and gave the rank order of dissimilarity judgments in the Table
3.5. The OR department decided to help the president's decision using the
nonmetric MDS technique.

Table 3.5   Rank order of dissimilarities between
             pairs of candidates*

A_i	1	2	3	4	5	6	7
1	—						
2	21	—					
3	13	10	—				
4	18	8	16	—			
5	3	19	4	20	—		
6	7	12	6	5	11	—	
7	17	1	2	14	15	9	—

* The rank number '1' represents the most similar pair.

Configuration of Alternatives: ALSCAL [374], a computer code for MDS
is used for the configuration. We start with two dimensions (t = 2) and
obtain the following configuration (see also Fig. 3.16);

A_i	D1M1	D1M2
1	1.6196	0.5867
2	-1.5199	-0.3735
3	0.2405	-0.9098
4	-1.1257	1.3917
5	1.5321	-0.4227
6	0.0452	0.6705
7	-0.7918	-0.9430

The stress of the above configuration is 0.2%, hence we do not have to
consider higher dimensionality.

Interpretation of Dimensions: We interpret the dimensions using a
multiple regression procedure to regress each attribute X_j over two dimensions (with 7 observations). The result is given in Table 3.6. Only three
attributes X_1, X_4 and X_5 give the significant regression. In other words
only these attributes affect the similarity judgments of the president.
The second and third columns of Table 3.6 show the optimum weights corresponding to each multiple correlation. (These are the direction cosines,
that is, regression coefficients normalized so that their sum of squares
equals 1). Among the significant regressions the significant weights are
again identified, for example X_1 has significant regression weight of
.9986. A regression weight of .9986 corresponds to an angle of 3.080
degrees from the dimension 1 since cosine (3.080) = .9986. Similarly
directions of X_4 and X_5 are shown in Fig. 3.16.

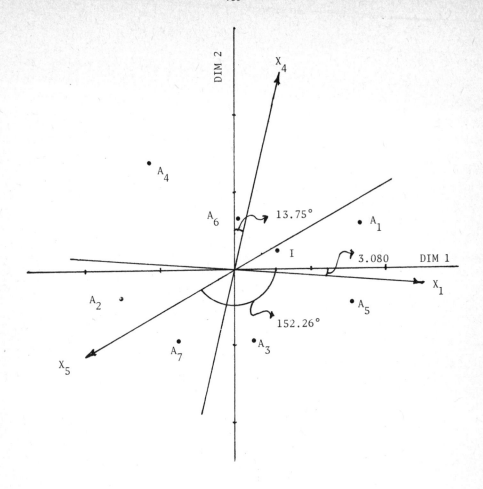

Fig. 3.16   Two-dimensional configuration by the ALSCAL program for
7 candidates.

Table 3.6   Multiple regression of each attribute
              on configuration dimensions[+]

X_j	Regression Weights (Direction Cosines)		Direction Degree		Multiple Corr.
	D1M1	D1M2	D1M1	D1M2	
1	.9986**	.0537	3.08°		.996*
2	.7410	.6715			.152
3	.9721	.2345			.820
4	-.2376	.9713**		13.75°	.989*
5	-.8851**	-.4654	152.26°		.984*
6	-.2869	-.9580			.302

 * Significant regression at .01 level ($F_{2,4}$).

** Significant coefficient at .01 level.

 [+] Result obtained by SAS computer code.

<u>Identification of Ideal Point</u>: LINMAP procedure (see Section 3.1.1)

may be utilized, but the president thinks he can locate his ideal point

easily in the two dimension space. The ideal point of the president is

(.5, .25). Then candidate A_6 (which has the shortest distance from the

ideal point) may go to Paris.

<u>Note</u>

This nonmetric MDS procedure suggests several points [273]. For one

thing, each pair of alternatives given in the judgment must be ranked in

terms of its similarity. This requires quite fine discrimination and may

be the most difficult for the DM among the MADM methods. Actually we arrange this

method at the end of the MADM list. The configuration is not unique. In

addition, the distance measure used to form the configuartion assumes that

the attributes (i.e., the dimensions of the chosen configuration) are independent

(i.e., noncomplementary). The attribute values (i.e., x_{ij}'s in the decision

matrix) can be in any form, since the scaling procedure itself produces

numerical, comparable values on each of t dimensions, $t \leq n$. It should also

be noted that none of the t dimensions necessarily correspond with single

attributes of the original decision matrix.

In spite of the greater burden, this method is particularly useful when

the number of attributes is quite large, say more than 7, and when the majority

of attributes are expressed in the qualitative way.

The drive to have practical models for management systems has been the impetus towards the development of many excellent methods for MADM methods presented in the preceding sections. However, few methods have yet been tested with real-world problems.

The most difficult task in the employment of MADM methods is determining the relevant set of attributes. This section aims to help planner, manager, researcher, policymaker, teacher, consumer, and so on to generate objectives/ attributes through the examination of the relevant literature.

The applications, which are classified according to the physical features of selected items instead of related area, are selection of commodity, site, people, project (plan, strategy), and public facility as shown in Table 4.1. Brief descriptions of the problem settings, list of attributes used, and MADM methods used are given in this section.

1.   COMMODITY SELECTION

To design new products and/or improve existing ones, manufacturers have a keen interest in obtaining preference information from consumers. Hence much of MADM study in the marketing area has been carried out to predict/explain consumer choice behavior rather than to help the individual consumer select among multiattribute market commodities. See Green and Srinivasan's review [154] for various choice models and Green et al's works [BM 8-10, 145-156] on this topic. Hammond's bullet selection [165] and Jacquet-Lagreze's magazine selection model [205] show how an individual/organization can improve choice decision.

Automobile Selection: Green et al. [BM-8, BM-10, 156] use nonmetric multidimensional scaling for consumer preferences when choosing among auto- mobiles.

Table 4.1   Classification of references on applications

	Class	References
1.	Commodity Selection	BM 8-9, 145, 156, 156a, 165, 205, 319 324
2.	Facility Location (Siting)	BM-13, 51a, 69b, 84, 157, 158, 181, 230 232, 274, 302
3.	People/Personnel Selection	71, 84, 91, 148, 186, 239, 297, 377
4.	Project Selection	
	4.1 Environmental Planning	79, 92a, 175, 239a, 267
	4.2 Land use planning	86, 87, 183, 184, 191, 264, 303, 305, 306, 315, 413
	4.3 R & D project	162, 178, 258, 268, 300, 309, 327, 361
	4.4 Water resources planning	67, 86, 160, 234, 292, 401
	4.5 Miscellaneous	BM-13, 16, 62, 85, 141, 174, 192, 225, 273, 318, 321, 406, 415, 426
5.	Public Facility Selection	BM-17, 108, 133, 164, 198, 199, 207, 243, 332, 350
6.	Miscellaneous	51, 65, 92, 133, 164, 231, 286, 362, 396

Park [319] reports usage of a simplified satisficing-plus model (a variation of conjunctive method) in the selection of automobiles. Ten selected attributes are: price, gas mileage, car size, car style, trunk space, frequency of repairs, safety features, durability of paints, car manufacturer, and number of doors.

Pekelman and Sen's study [324] for the evaluation of automobiles is to assess the weight of attributes using LINMAP technique. Seven attributes are considered: styling, warranty, comfort, dependability, acceleration, purchase cost, and youthfulness.

Bullet Selection: Hammond [165] studies the case of selecting bullets for the police department in Denver. Three characteristics of ammunition are considered: stopping effectiveness, amount of injury, and threat to bystanders as a result of penetration of the body and/or ricochet.

Computer System Selection: Grochow [156a] employs a multiattribute utility model in the evaluation of computer systems. The attributes considered are: response time to trivial requests, response time to computer-bound requests, availability, and reliability.

Food Selection: Green et al. [145, 156] use nonmetric multidimensional scaling and conjoint measurement (a kind of regression model) models in order to analyze consumer preferences for choosing food items.

Pekelman and Sen [324] use LINMAP model to assess the weight of attributes in the selection of dry cereal. Six attributes considered are: crispiness, crunchiness, sweetness, fillingness, calorie content, and flavor.

Magazine Selection: Jacquet-Lagreze [205] studies the problem of selecting magazines to be used in an advertising campaign. He utilizes the ELECTRE model to evaluate 16 magazines under 6 attributes: editorial content, price of space, power (representing the importance of the number of readers and reliability of the information given), affinity of magazine, and prestige of magazine.

2.   FACILITY LOCATION (SITING) SELECTION

The siting of (public) facilities is a complex task and needs extremely
discreet decision making. For instance decision making for the siting
of nuclear facilities involves many concerned interest groups, huge construction cost, and impact on the surrounding area (possibilities of severe
environmental damage), etc.

Browning's book [41b] on this topic will help decision makers in
the generation of relevant attributes.

Site for Airport: Charnetski [51a] reports the case of selecting
one of three proposed sites for a modern air terminal system in south Florida.
He generates 52 attributes under three major categories: ecological and
environmental, sociological and economic, and systematic-performance measures.

Site for (Nuclear) Power Plant: A study to identify and recommend
potential new sites in the Pacific Northwest suitable for a thermal electric
power generating station is made by Keeney and Nair [232]. They initially
screen the alternatives by conjunctive method and evaluate the remaining
alternatives by the multiattribute utility function (multiplicative form).
Criteria used in the screen process are: health and safety (measured by
distance from populated areas, height above nearest water source for flood
consideration, and distance from fault), environmental effects (measured
by thermal pollution, and location with respect to ecological areas), socio/
economic effects (measured by location with respect to designated scenic
and recreational areas),and system cost and reliability (measured by cost
of cooling water acquisition, cost of pumping water, and cost of providing
access for major plant components). The attributes used in evaluating
the candidate sites are: site population factor (relative human radiational
hazard associated with a nuclear facility), loss of salmonids, biological
impacts at site, length of transmission intertie line through environmentally

sensitive areas, socioeconomic impact, and annual differential cost between sites.

A similar study was made by Gros et al. [157] for the nuclear facility siting along the New England coast. The utility functions for four interested groups are assessed over four attributes: number of units at a site, cost, population within ten kilometers of a site, and incremental water temperature at peak ambient water temperature period of year.

Hill and Alterman [181], and Davos et al. [69b] report the case study of power plant site evaluation for Israel and the State of California, respectively.

Site for Pumped Storage Facility: Keeney [230] evaluates 10 sites for the pumped storage hydroelectric generation facility. The attributes in his multiattribute utility are: first year cost, transmission line distance, forest lost, and community lost due to the construction.

Site for Recreation Area: Kahne [274] uses simple additive weighting model (allowing uncertainty in attribute) to determine the relative quality of over 150 outdoor recreation areas in the State of Minnesota. Twenty-nine attributes are considered.

Site for Residential Environmental: Nakayama et al. [302] use multi-attribute utility function (multiplicative form) to assess the residential environment of Kyoto in Japan. Twelve attributes are considered in the utility function representing "good residential district": proportion of green area, proportion of park area, population density, medical facilities bad smell, traffic accidents, sulphurous acid gas, soot and smoke, factories, accessibility to the center of the city, offices of the business affecting public morals, and land price.

3. PEOPLE/PERSONNEL SELECTION

A basic problem facing all organizations is that of selecting qualified people to work for them. Employment of MADM analyses in this field are beneficial both to individuals and to the organizations.

Selection of All-Star Basketball Players: Einhorn and McCoach [91] use multiattribute utility function (simple additive form) to evaluate the player performance in the National Basketball Association. Eight attributes of player performance are used: field goal percentage, free throw percentages, rebounds, assists, steals, personal fouls, points per minute played, and blocked shots. The model yielded a rank order of all forwards, guards, and centers for each season. This ordering then is compared with the NBA all-star team. The results showed that the model predicted the all-star team very well.

Personnel Selection: Smith and Greenlaw [377] report a simulation model for the hiring of company employees.

Easton [84] compares three evaluation rules (geometric mean, arithmetic mean, and quadratic mean) for the selection of sales manager.

Selection of Professorial Candidates for Tenured Positions: Green and Carmone [148] make a pilot study involving graduate business students' evaluation of (hypothetical) assistant professors in terms of their suitability for tenured positions. A regression model with three criteria (research and publication, teaching, and institutional contribution) is utilized.

Selection of Entering Students: Hirschberg [186] reviews graduate student selection policies and reports that a simple linear regression model is extremely robust.

Dawes [71] represents the judgments of a university committee admitting

PhD students. The attributes of prime importance are test scores on standard
exams (GRE), the student's undergraduate grades (GPA), and the quality
of the undergraduate school attended (QI). The linear paramorphic model
is: .0032 GRE + 1.02 GPA + .0791 QI.

Moscarola [297] uses ELECTRE model for the selection of candidates
for business school admission. The attributes considered are: high school
grade average, improvement, experience, motivation, professional interest.

Klahr [239] applies spatial representation (using nonmetric multidimensional scaling with ideal point) for the prediction of college admission
officers' preferences for choosing among student applicants. The initial
set of attributes considered contains: alumni interview, campus interview,
college board scores, extracurricular activities, high school grade average,
high school recommendation, IQ, and rank in senior class.

4. PROJECT SELECTION

There exist multiple alternatives in any planning stage. A proper
evaluation of plan and strategy on project alternatives is essential for
the successful implementation. The fields of project selection are too
diverse to classify them systematically. However, they are divided into
five groups according to the number of reported cases.

4.1 Environmental Control Planning

Air Pollution Control: Ellis and Keeney [92a] report use of a multiattribute utility model in the evaluation of two air pollution control
strategies for New York City. These alternatives are the status quo, which
entails maintaining a one percent legal limit on the sulfur content of
oil and coal used in New York City, and one that lowered the legal limit
to 0.37 percent for oil and .7 percent for coal. Seven attributes considered
are: per capita increase in the number of days of remaining lifetime,
per capita decrease in the number of days of bed disability per year, per

capita annual net costs to low-income residents, per capita annual net costs to other residents, daily sulfur-dioxide concentrations in parts per million, total annual net cost to city government, subjective index of political desirability.

Environmental Index: Dinkel and Erickson [79] suggest an "Environmental Index" for the evaluation of environmental program effectiveness.   The criteria of this index are: number of serious pollution incidents, number of less serious pollution incidents, number of complaints, comparison of environmental quality, compliance index, and number of nonmonitored industries by type, size, and region.

Sewage Sludge Treatment: Herson [175] evaluates land-based alternatives for the disposal of large quantities of sewage sludge generated in a city treatment plant of Los Angeles.   The impacts (advantages and disadvantages) of the three alternatives (agricultural, evaporation pond, and landfilling) are supplied instead of decision matrix form.

Solid Waste Management Problem: One real problem faced by the Bureau of Solid Waste Management involves the dismantling of retired wooden railroad cars.   Three alternatives for wood removal in salvaging the metal from retired railroad cars are: open burning, grit-blasting, and use of a high pressure water jet.   Klee [239a] suggests DARE technique (a variation of hierarchical additive weighting) for the evaluation of these alternatives under five attributes: capital cost of the facility, ability of the process to salvage the wood removed, time needed to develop the process, contribution to air pollution of the process, and operating cost of facility.

Litchfield et al. [267] consider an analysis of a hypothetical advanced nuclear waste management system by multiattribute utility theory.

4.2 Land Use Planning

Land use planning may be defined as a process for determining the optimal allocation of resources in order to achieve a set of objectives in space. This problem is especially important for land poor countries such as the Netherlands and Israel.

Land Reclamation Project: Nijkamp et al. [303, 305, 306, 413] present an ELECTRE model for a land reclamation project in the Netherlands. The attributes considered are additional natural area, additional agricultural area, additional recreational area, residential opportunities, increased accessibility, additional employment, relative importance of new airport, annoyance of a new airport, investment cost, etc.

Paelinck [315] uses permutation model for a similar project. Hill and Tazmir [183] use multidimensional scalogram analysis for the case of Israel.

Coastal Zone Planning: Edwards [86, 87] employs multiattribute utility (simple additive form) to evaluate coastal zone planning in California. The attributes considered are: size of development, distance from the mean high tide line, density of the proposed development, onsite parking facilities, building heights, unit rental, conformity with land use in the vicinity, esthetics of the development.

Selection of Load (Pipe) Line: Lhoas [264] uses ELECTRE model for the selection of the route for a pipeline designed to supply a group of refineries with crude oil in Belgium. His evaluation criteria include: total financial cost, period over which the solution is expected to be valid, possibilities of simultaneously laying down a pipe network for the transport of other products, possibilities of connecting refineries with

various supply ports, priority of supply, interdependence of the countries
involved, regional independence, development possibilities of Belgian harbor
activities, etc.

Holmes [191] employs an ordinal model (a primitive form of linear
assignment method) for the evaluation of four alternatives for a new load
line. Twenty-seven attributes are identified.

4.3 Research and Development (R & D) Project

Research organizations have a tendency to propose problems for investigation at a faster rate than resources can be supplied to support the
work. This poses a fundamental issue for the managers of (industrial)
research because there are always more ideas and problems to be investigated
than available resources will permit. Research management, therefore, is
faced with the problem of selecting a set of projects which will achieve
maximum effectiveness for the organization [300].

Mottley and Newton [300] consider the selection of industrial research
projects leading to the development of new products, processes, and uses.
Their utility model has five attributes: promise of success, time to completion,
cost of the project, the strategic need, and market gain.

Gustafson et al. [162] use simple additive weighting model in the
evaluation of R & D projects. They use hierarchical structure for assessing
attribute weights under four major categories: benefit, regional considerations,
methodology, and capacity.

Other studies for this area are: aerospace industry for military
and commercial system planning by Heuston and Ogawa [178], information
system development by Lucas et al. [268], general R & D selection by Larichev
[258], Nowlan [309], Pinkel [327], and Schwartz et al. [361].

4.4 Water Resources Planning

David and Duckstein [67] use ELECTRE model for evaluating five alternative long-range water resources development plans for the Tisza River basin in Hungary. Twelve attributes considered are: costs, probability of water shortage, energy (reuse factor), recreation, flood protection, land and forest use, manpower impact, international cooperation, development possibility, flexibility. The same problem is treated by Keeney and Wood [234] by the assessment of multiattribute utility function (multiplicative form).

Minnehan [292] employs a series of MADM methods (dominance, conjunctive method and simple additive weighting) for the choice of water resources project design in the Sacramento-San Joaquin Delta in California.

4.5 Miscellaneous

Academic Planning: Hopkins et al. [192] construct a multiattribute attribute utility function (simple additive form) that reflects the preferences of university administrators for different university configurations. This function is to be used in a mathematical programming model to optimize the size and shape of the institution. LINMAP is utilized to obtain the weight for each attribute. The attributes considered are: regular faculty, auxiliary faculty, undergraduate enrollment, graduate enrollment, graduate enrollment in professional programs, tuition level, staff/faculty ratio, and faculty leverage.

Air Defense System Development: Eckenrode [85] suggests six attributes to be considered for air defense system development. They are: economy, early availability, lethality, reliability, mobility, and troop safety.

MacCrimmon's review [273] illustrates the choice of future missile systems through various MADM methods. Six attributes considered are: range, delivery time, total yield, accuracy, vulnerability, and payload delivery flexibility.

Airport Development Planning: Keeney [BM-13, 225] reports an analysis done for the Mexico government to help select the most effective strategy for developing the airport facilities in the Mexico City metropolitan area to insure quality air service for the remainder of the century. A multi-attribute utility function (multiplicative form) is utilized with six attributes: total cost, practical capacity (number of aircraft operations per hour), access time to and from the airport, number of people seriously injured or killed per aircraft accident, number of people displaced by airport development, and number of people subjected to a high noise level.

Capital Investment Planning: Gearing et al. [141] measure the tourist attractiveness for allocating the capital investment and selecting tourist projects for the Turkish Ministry of Tourism. Touristic attractiveness contains 17 attributes under five major categories: natural factors, social factors, historical factors, recreational and shopping facilities, and infrastructure and food and shelter.

Wehrung et al. [426] use the dominance model in order to explain the investment preferences for over 400 top level business executives in Canada and U.S. They report the dominance model with two criteria (expected rate of return and variance) fits well for their selection behavior among other criteria, e.g. probability of specific gains, probability of specific losses, negative semivariance, etc.

Hax and Wiig [174] use decision tree technique with multiattribute utility function in the selection of a capital investment project.

Health-care facility planning: Parker and Srinivasan [321] use LINMAP model for the patients' overall preferences in the expansion of a rural primary health-care delivery system. The facility attributes considered are: residence-to-facility travel time, time to get an appointment, waiting time at the facility, facility hours of operation, type of health care facility, provider of medical care, range of services and technical facilities available, and price per visit.

Multiattribute utility functions are formulated for the aggregate scheduling in health maintenance organizations by Vargas et al.[415].

Joint Venture Strategy in Fisheries: Tomlinson and Vertinsky [406] study the evaluation of alternative joint venture strategies for exploitation of fisheries in the economic zones of the oceans. Five alternatives are generated: null strategy, licensing strategy, foreign subsidiaries strategy, joint venture strategy, and independent strategy. Delphi method is utilized to reach consensus of strategic choices among coastal countries.

Forest Pest Control Problem: The forests of New Brunswick, Canada have frequently been attacked by an insect pest known as the spruce budworm. Bell [16] assesses a multiattribute utility function reflecting the preferences of one member of a simulation project for the condition of the forest and economy. The attributes are profit, unemployment, and recreational value of the forest.

Transportation System Effectiveness: Pardee et al. [318] study the transportation system effectiveness for the Northeast Corridor Transportation project. They list numerous attributes under three major affected groups using the hierarchical structure of attributes. A simplified version of this study is done by Crawford [62].

5. PUBLIC FACILITY SELECTION

Marketing research has given less attention to the selection of public facilities (like banks, hospitals, companies, etc) compared with that of commodities. But this area is also important to improve the existing systems/ facilities by assessing consumer's preferences as well as to improve each individual's decision making.

Bank Selection: Consumer preferences for banks in which to open a checking account is studied by Jain et al. [207]. They utilize LINMAP, regression model, etc. using five attributes: cost of checking account, type of bank, accessibility to banking service, quality of service, and banking hours.

Hospital evaluation: The evaluation study for the quality of twelve hypothetical hospital wards is performed by Huber et al. [199]. Seven variables considered in their regression models are: total nursing hours of care per patient per day, nurse-patient and nurse-physician communication, percent professional nursing care, neatness, cleaniness and orderliness, head nurse administrative capability, number of beds in ward, and age of ward.

Job Selection: The problem of selecting the best job or profession after college graduation is treated by Fishburn [108], Huber et al. [198], Miller [BM-17]. Miller's utility function (simple additive form) considers 15 attributes under four major categories: monetary compensation, geographical location, travel requirement, and nature of work.

School Selection: Saaty [350] uses hierarchical additive weighting model for the evaluation of high schools. His selection criteria are: learning, friends, school life, vocational training, college preparation, and music classes.

The college choice behavior of graduate business school applicants is studied by Punz and Staelin [332]. Results indicate that such factors as price (tuition), fellowships, distance from home, and quality of school are relevant in their selection.

Kohn et al. [243] also study   college-going behavior. Their utility model considers cost, academic quality, quality of life, etc.

Miscellaneous: Foerster [133] makes a comparative study among some MADM methods for the prediction of ground transportation system choice behavior. He reports that conjunctive, lexicographic, and conjunctive-lexicographic methods outperform (commonly used) simple additive weighting.

A study on consumer preferences for hypothetical restaurants is performed by Hagerty [164].

V. CONCLUDING REMARKS

This survey of MADM methods and applications is a sequel to our previous survey of MODM methods and applications [BM-11]. It is a guided tour through the literature on the subject. It provides readers with a capsule look into the existing methods, their characteristics, and their applicability to analysis of MADM problems.

On MADM Methods Classification:

We have proposed a system of classifying seventeen major MADM methods (see Fig. 1.1 A taxonomy of methods for multiple attribute decision making); half of them are classical methods in this field, and the other half have been proposed recently by various researchers in diversified disciplines, but here for the first time they are presented together. The literature of these methods is identified and classified systematically (see Table 1.3).

On Applications of MADM:

MADM applications are classified into several categories. Each topic's references are identified (see Table 4.1), and a summary of each article is presented. The MADM models are presented in a diversified field of applications, but reported examples of applications of MADM methods to real-world problems seem to be rare.

On Multiple Objective Decision Making (MODM) Methods:

In the study of decision making in complex environments, such terms as "multiple objective," "multiple attribute," "multiple criteria," or "multiple dimensional" are used to describe decision situations. Often these terms are used interchangeably, and there are no universal definitions of these terms [BM-13]. Multiple criteria decision making (MCDM) has become the accepted designation for all methodologies dealing with MODM and/or MADM.

Our definitions distinguish between MODM and MADM (see II Basic Concepts and Foundations). In our survey the problems of MCDM are broadly classified into two categories - MADM and MODM. MADM methods are for selecting an alternative from a small, explicit list of alternatives, while MODM methods are used for an infinite set of alternatives implicitly defined by the constraints [275]. MODM methods thus address design problems while MADM methods are useful in choice problems.

Literature on MODM methods and applications has been reviewed extensively by us in [BM-11]. MODM methods were systematically classified; a taxonomy of the methods is presented in Fig. 5.1.

On Multiattribute Utility Theory (MAUT):

In this review we have so far focused on riskless choice models that involve multiattribute alternatives. We have avoided the added complexity of uncertainty. Uncertainty can arise in two ways: (1) as uncertainty about the outcome, or (2) as uncertainty about the attribute values.

Multiattribute utility theory (MAUT) has been developed for the uncertainty about the outcome (consequences). If an appropriate utility is assigned to each possible consequence and the expected utility of each alternative is calculated, then the best course of action is to take the alternative with the highest expected utility. But the assessment of appropriate multiattribute utility function is so complex and intricate that the aim of the most theoretical work in MAUT is the investigation of possibilities for simplifying multiattribute utility assessment procedures. More recent advances in MAUT have been developed principally by Keeney [219, 220, 222, 224, 226] among others. The literature on MAUT and its assessment methods has been summarized in Farquhar [98], Fischer [105, 107], Fishburn [120, 126], Huber [196, 197], and Winterfeldt and Fischer [434]. Keeney and Raiffa [BM-13] in particular deal extensively with utility

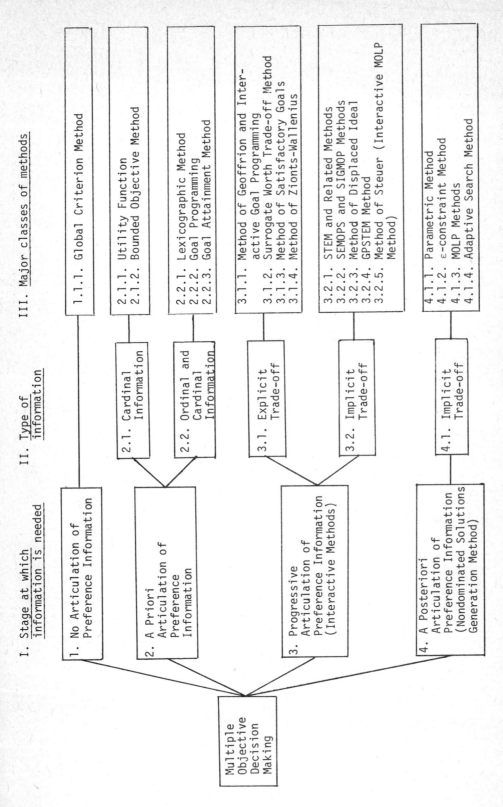

Fig. 5.1 A taxonomy of methods for multiple objective decision making [BM-11].

theory from unidimensional to multiattribute, its assessment methods and applications.

A Choice Rule for MADM Methods:

As to which method(s) we should use, the selection of MADM method(s) itself is a MADM problem. There exists no specific choice rule. Different MADM methods are introduced for different decision situations. We suggest a general choice rule represented by a tree diagram in Fig. 5.2. The numbers in the circles refer to the MADM methods in the taxonomy (see Fig. 1.1). The proper method(s) emerge(s) by answering the subsequent questions.

The first question in Fig. 5.2 asks the DM whether he is trying to find the best alternative or will take any of the alternatives which satisfy the minimum acceptable level(s) of attributes. When the DM does not have enough time and knowledge to examine the problem further the answer to question 1 may be "no" and he gives the minimum acceptable level(s) for attribute(s).

If the minimum levels for multiple attributes are given, the method of conjunctive constraints (2.1.1) is used; and if the minimum level for one attribute is given, the method of disjunctive constraint (2.1.2) is used.

If the answer to the first question has been "yes", the DM wants to get a better solution with the extra effort. Then different followup questions are asked.

The second question asks whether the dominated alternatives are screened. If the answer is "no", they are eliminated by dominance (1.1.1). If the answer is "yes", then different followup questions are asked.

In order to use any MADM methods presented in the taxonomy, it is required that there be a single DM or that the multiple decision makers should develop aggregation schemes for assessing group decisions. The

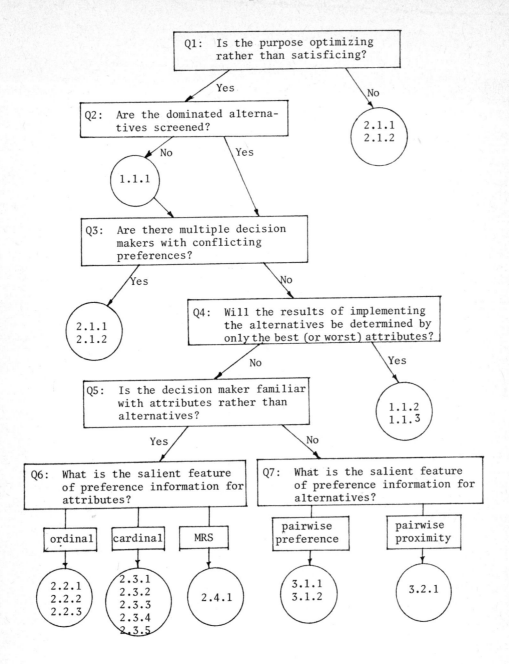

Fig. 5.2.   MADM method specification chart.

third question asks whether the group of decision makers reaches the agreement in their preferences. If they can not, the easiest way with the multiple decision makers is to impose different constraints and solve the problem by conjunctive constraints (2.1.1) or by disjunctive constraint (2.1.2).

If MADM methods having conflicting preferences are not being dealt with, it must be ascertained (in question 4) whether performance will be determined by the best or the worst attribute. If the answer to the question is "yes", and if the best attribute is the determinant, maximax (1.1.3) can be used; if the worst attribute is the determinant, maximin (1.1.2) can be used. If the answer is "no", proceed to question 5.

The fifth question asks whether the DM is familiar with attributes rather than alternatives. It depends on the ability   the DM can assess the preference by attributes or by alternatives.   It needs empirical judgment. Usually as the number of attributes is increased, say to more than seven, preference assessment by alternatives becomes extremely difficult. If the DM is more familiar with attributes, he will supply the attribute preferences and answer question 6, otherwise he will answer question 7.

The DM is now to decide which kind of preference inputs seems most valid (in question 6 and 7). To help answer question 6, consider the preference inputs in roughly the order of the least demanding to the most demanding. If the interattribute ranking (ordinal) can be assigned, we can use lexicographic method (2.2.1), elimination by aspects (2.2.2), or permutation method (2.2.3). If the interattribute numerical weights (cardinal) can be assigned, then the linear assignment method (2.3.1), simple additive weighting (2.3.2), hierarchical additive weighting (2.3.3), ELECTRE (2.3.4), or TOPSIS (2.3.5) can be utilized. Even if such numerical information is not meaningful, the DM may be able to make some numerical tradeoffs (marginal rate of substitution). The method of hierarchical tradeoffs (2.4.1) is used. ELECTRE (2.3.4) is believed to be a most refined method in this class.

If the pairwise preferences for the alternatives are assessed, we can use LINMAP (3.1.1) or interactive simple additive weighting (3.1.2). If the order of pairwise proximity is assigned, nonmetric multidimensional scaling with ideal point (3.2.1) can be used. The assessment of order of pairwise proximity is the most demanding preference input.

A Unified Approach to MADM:

Behavioral scientists, economists, and decision theorists have proposed a variety of models/methods describing how a DM might arrive at a preference judgment when choosing among multiple attribute alternatives. Different MADM methods involve various types of underlying assumptions, information requirements from a DM, and evaluation principles.

Some methods may not fully satisfy the underlying assumptions (e.g., complete independence among attributes for simple additive weighting method), and some may have more tedious procedures (e.g., constructing indifference curves for hierarchical tradeoffs than others). There are also methods fitted for a specific situation only (e.g., maximin for the pessimistic decision situation, maximax for the optimistic, disjunctive method for the selection of specialized person). Even with the same preference information, each method evaluates the problem from a different point of view (e.g., with a set of weights, weighted average outcomes for simple additive weighting, relative closeness to the ideal solution for TOPSIS, satisfaction to a given concordance measure for linear assignment method and ELECTRE).

How, then, does a DM select an appropriate MADM method in general situations? If the DM already has a fixed form of his/her preference information, one may narrow down his/her options by following the branches of MADM taxonomy in Fig. 1.1. Otherwise he/she still has to select one method. The existence of multiple MADM methods seems to be more burdensome than helpful to those who want to use one method. Easton [84] mentioned that "the selection of a MADM method is itself a MADM problem."

An attempt [440a] is made to reach a unified solution by employing some MADM methods simultaneously instead of selecting the best method for the situation. The proposed approach consists of four phases whose main features are (see Figure 5.3):

Screening phase: The dominated and unacceptable/infeasible alternatives are eliminated by dominance and conjunctive method.

Preference ordering phase: Four efficient MADM methods, linear assignment method, simple additive weighting method, TOPSIS, and ELECTRE, using cardinal preference information, evaluate the decision problem independently.

Aggregation phase: Three ordering techniques--statistic, Borda method and Copeland method--will be applied to aggregate the four sets of preference rankings obtained in the previous phase.

Synthesis phase: This phase attempts to reach consensus among three ordering techniques. The consensus is made by a partial ordering technique [79a] which synthesizes these different viewpoints.

Screening Phase:

This phase tries to cut down on the information processing requirements by eliminating dominated alternatives and removing unacceptable alternatives before entering the evaluation phase. Dominance (1.1.1) and conjunctive method (2.1.1) will be utilized in this phase. Any alternatives passing these two sieves become the candidates for further evaluation.

Eliminating dominated alternatives: An alternative is dominated if there is another alternative which excels it in one or more attributes and equals it in the remainder. The number of alternatives can be reduced by eliminating the dominated alternatives. Interest in the nondominated set may be either as a sufficient indication of near-optimal choices or as a reduction in the number of alternatives to which it is necessary to apply more elaborate analytical techniques of MADM. Calpine and Golding [44] demonstrated convenience and effectiveness by using this sieve.

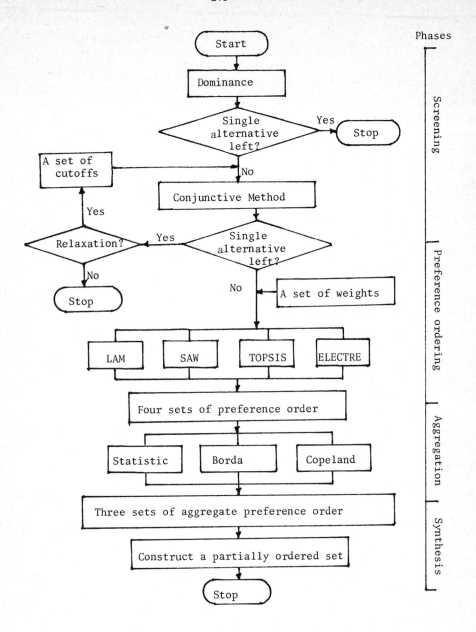

Fig. 5.3.   A process diagram for unified approach to multiple
attribute decision making.

Removing the unacceptable alternatives: A set of nondominated alternatives may possess unacceptable or infeasible attribute values. The conjunctive method is employed to remove the unacceptable alternatives. This step tries to dichotomize alternatives into acceptable/not acceptable categories rather than to evaluate the alternatives; hence it is advisable not to set the cutoffs high. Normally multiple alternatives are expected to remain after the conjunctive screening. If none or a single alternative is left, the relaxation of the cutoffs is subject to the DM.

Preference Ordering Phase:

This phase tries to find a set of preference orders of alternatives by employing some MADM methods simultaneously. Four efficient MADM methods utilizing a weighting scheme are selected. They are linear assignment method (2.3.1), simple additive weighting (2.3.2), TOPSIS (2.3.5), and ELECTRE (2.3.4).

Aggregation Phase:

Three ordering techniques will be applied to aggregate the four sets of preference orders. The first technique ranks alternatives according to their mean rankings. The second (Borda method) is based on a majority rule binary relation. The last procedure (Copeland method) is a modification of the majority rule case taking into account "losses" as well as "wins".

The Borda and Copeland procedures are described in detail elsewhere [94, 116a]. Therefore a simple example is given to illustrate the procedures and characteristics of each technique. Assume that four alternatives are ranked by the four MADM methods in the previous stage. Their results are shown in Table 5.1. This data will be aggregated by the three techniques.

Average ranking procedure: The average rank order (over four MADM methods) is shown in the column labeled "average". Therefore, based on the average rank, the aggregate rank order is A_2, A_1, A_3, and A_4. If two alternatives have the same average rank, the one with the smallest standard deviation can be ranked ahead.

Borda method: It is based on a majority rule relation. Given the example data of Table 5.1, one can determine the set of MADM methods which prefers A_1 to A_3; in this case, it is $\{M_1, M_2, M_3\}$, i.e., three methods. Similarly one can see that only one method prefers A_3 to A_1, i.e., M_4. Therefore, the number of methods preferring A_1 to A_3 is greater than the number of methods preferring A_3 to A_1; hence A_1 has a majority vote over A_3, denoted $A_1 \ M \ A_3$. Because an even number (= four) of methods is considered in the previous stage, a tie may be possible as in the case of A_1 and A_2. They are denoted as $A_1 \ T \ A_2$.

Each pair of alternatives has to be considered separately. This implies $m(m-1)/2$ determinations where m is the number of alternatives. The results of this process are shown in Table 5.2. The ordering criterion is based on how many times an alternative "wins" a majority vote. The last column in Table 5.2, labeled ΣC represents the number of times an alternative "wins" in the voting. The alternative with the highest sum for ΣC is ranked first, etc. Ties can again occur in that several alternatives have the same sum. These alternatives are then identified as a group and ranked at the same level. The aggregate rank of the example is (A_1, A_2), A_3, and A_4.

Copeland method : This procedure starts where the Borda method stops. The Copeland method considers not only how many "wins" an alternative has, but also explicity includes the "losses". The bottom row of Table 5.2 represents the number of "losses" for an alternative. For example, alternative A_3 lost to both A_1 and A_2. Therefore, reading down column A_3 one finds two M's. The Copeland score is determined by substracting the "losses" for an alternative from its "wins". For example, the score of A_3 is $(1 - 2) = -1$. The scores for A_1 through A_4 and 2, 2, -1, and -3, respectively. This implies an ordering of (A_1, A_2), A_3, and A_4. Since the number of "wins", "losses", and "ties" must add up to $(m-1)$, not only "wins" but also "losses" and "ties" are considered in the Copeland method.

Table 5.1. Preference rankings by four MADM methods

	MADM Methods				Average
	M_1	M_2	M_3	M_4	
A_1	1	2	1	4	2
A_2	2	1	2	1	1.5
A_3	3	3	4	2	3
A_4	4	4	3	3	3.5

Table 5.2. Majority vote between alternatives

	A_1	A_2	A_3	A_4	ΣC
A_1	—	X	M*	M	2
A_2	X	—	M	M	2
A_3	X	X	—	M	1
A_4	X	X	X	—	0
ΣR	0	0	2	3	

* An M in the i, j cell implies that $(A_i \, M \, A_j)$, which also implies that an X appears in the i, j cell, implying not $(A_i \, M \, A_j)$.

From the above observation one can see that each ordering strategy views the original rank-ordered data of the four MADM methods from different perspectives. This represents a rich source of analytical information if it can be meaningfully synthesized. The next phase presents a method for synthesizing all this information.

Synthesis Phase:

This phase attempts to reach consensus among three ordering strategies through the construction of a partially ordered set (Poset) [79a, 94]. The procedure is based on a method described by Dushnik and Miller [79a] whereby a set of linear orders can be aggregated to form a poset. For example, if one has the set of orderings

$$K = \{0_1, 0_2\}$$

where

$$0_1 \text{ is } A_1 > A_2 > A_3 > A_4$$

and

$$0_2 \text{ is } A_2 > A_1 > A_3 > A_4$$

and all the elements of 0_1 and 0_2 come from the same set, i.e., $S = \{A_1, A_2, A_3, A_4\}$, then, the partial order P_1 can be formed:

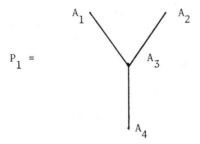

$$P_1 =$$

where A_4 is always less preferable to any other element, and A_3 is always less preferable to A_1 and A_2. The elements A_1 and A_2 are not comparable in P_1 since in 0_1, $A_1 > A_2$ and in 0_2, $A_2 > A_1$.

Therefore, in constructing a partially ordered set P from a collection
of linear orderings K (all the orders contain elements from a common
set S), one uses the following procedure:

For any pair of elements x_1, x_2 in S, $x_1 < x_2$ in P, if and only
if $x_1 < x_2$ in every linear order of K. A partial order so obtained will
be said to be realized by the linear orders of K.

This method does not need to be modified in order to handle the tie
cases. The same rule applies: only in the case of ties (between x_1 and
x_2) it is not true that $x_1 < x_2$ and it is not true that $x_2 < x_1$, so they
are deemed incomparable.

The poset of the illustrative example in the aggregation phase has
the same structure as P_1 above.

It is important to note that the poset construction preserves the
uniqueness of each component ordering. One attribute is placed above another
if and only if it is above it in all the component orderings. Therefore,
the similarities of the component orderings are preserved and highlighted.
In fact, if all the component orderings are identical, the resultant poset
is an exact reproduction of the component ordering.

A Hypothetical Example:

A MADM problem with eight attributes is considered. After the screening
phase the number of alternatives is reduced to seven. Table 5.3 shows
the results of the preference ordering phase.

The mean rank order (over four MADM methods) is given in the last
column of Table 5.3. The results of majority vote are given in Table 5.4.
Aggregate ranking by Borda method and Copeland methods can be obtained
by reading the descending order of ΣC_i and ($\Sigma C_i - \Sigma R_i$) of Table 5.4, respectively.
The three sets of aggregate preference rankings are given in Table 5.5.

Table 5.3. Preference rankings by four MADM methods.

	MADM methods				AVE
	LAM	SAW	TOPSIS	ELECTRE	
A_1	3	7	4	4	4.5
A_2	4	5	6	5	5.
A_3	7	1	3	3	3.5
A_4	1	2	1	6	2.5
A_5	6	6	5	7	6.
A_6	5	4	7	1	4.25
A_7	2	3	2	2	2.25

Table 5.4. Majority vote between alternatives.

	A_1	A_2	A_3	A_4	A_5	A_6	A_7	ΣC
A_1	—	M	X	X	M	X	X	2
A_2	X	—	X	X	M	X	X	1
A_3	M	M	—	X	M	X	X	3
A_4	M	M	X	—	M	M	M	5
A_5	X	X	X	X	—	X	X	0
A_6	X	X	X	X	M	—	X	1
A_7	M	M	M	X	.M	M	—	5
ΣR	3	4	1	0	6	2	1	

Fig. 5.4 shows the partially ordered set realized by the orderings of Table 5.5. The poset is composed of four levels where there is disagreement among the alternative orderings. The poset reflects this by making the associated alternatives incomparable (i.e., tied relationship), e.g., A_4 and A_7 in level 4 and A_1, A_2, and A_6 in level 2. This is the final preference ranking which a DM expects to get by supplying a set of weights on attributes.

On Future Study:

Uncertainty in attribute values: In our review we have assumed that each attribute value was known, and that that value was unique. But we recognize that the information available to the DM is often highly uncertain, especially in research and development decision making. There are various ways of representing the decision maker's uncertainty. The simplest way is to use expected values for each attribute value and then treat the problem as one of certainty choice. A second and more computationally demanding procedure is to use an interval or range of values rather than a point estimate of attribute values. Some MADM methods such as dominance, disjunctive, conjunctive, and lexicographic method may somehow be modified to treat problems with uncertainty in attribute values, but the extension to other methods becomes computationally too cumbersome to be effective. A third and most complex way to account for attribute value with uncertainty is by introducing probability distribution. A recent approach to apply fuzzy set theory [217, 453] to MADM methods aims to overcome these difficulties. Bellman and Zadeh [19] have shown its applicability to MCDM study, and Yager and Basson [440], and Bass and Kwakernaak [8, 257] introduce maximin and simple additive weighting method using membership function of the fuzzy set. Many efficient MADM methods are waiting for accommodation to the attribute value uncertainty.

Table 5.5. Aggregate preference ranking from three techniques.

Rank	Average	Borda	Copeland
1st	A_7	A_4, A_7	A_4
2nd	A_4		A_7
3rd	A_3	A_3	A_3
4th	A_6	A_1	A_1, A_6
5th	A_1	A_2, A_6	
6th	A_2		A_2
7th	A_5	A_5	A_5

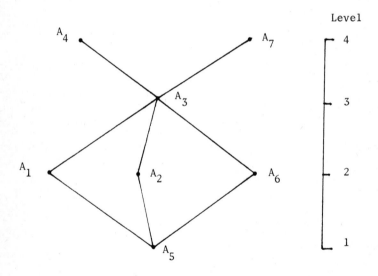

Fig. 5.4. The partially ordered set of Table 5.5.

Uncertainty in outcomes (consequences): The multiattribute utility theory (MAUT) has been developed for handling uncertainty in outcomes. But MAUT is one of the most confusing topics of MADM study mainly due to its sophisticated nature (assumption) and the difficulty in the assessment of utility function. Furthermore, most of the literature is filled with mathematical proofs. Though the aim of most theoretical work in MAUT is the investigation of possibilities for simplifying the task of MAUT assessment, there remains a certain amount of skepticism concerning the practical usefulness of MAUT. Research efforts should be directed to the user oriented transformation of MAUT. A busy DM needs a MAUT assessment technique which can be easily taught and used, though it may lack the theoretical elegance of techniques proposed by, for example, Keeney and Raiffa [BM-13].

Multiple decision makers: Throughout this study it is assumed that there is a single DM (or a group with 'one voice') who articulates preferences. One area which seems to be very important is multiple decision makers or group decisionmaking. Conjunctive method (satisficing method) may be directly applied in this situation if each DM imposes different constraints. And it is possible to use most MADM methods if group consensus can be obtained about the preference information. For instance, the preference information on attribute weight can be merged in some ways: the arithmetic mean [56], and weighted average depending upon the individual's power in a group [305,365a]. But each DM may have conflicting opinions as to (1) selection of decision criteria (attributes), (2) determination of attribute weight, (3) selection of evaluation method, etc. The group problem certainly has difficulties not presented in the single DM problem. Tell [399] proposes the revised Delphi procedures to handle these compounded difficulties. However, not much has been done in the way of developing methodology for approaching the group problem [64, 302, 399, 407]. It is hoped that this area will get more attention from researchers in future.

The development of methods for the "fuzzy" area which combines MADM and MODM problems may be a topic for future research. Many multiple criteria decision making problems blend selection (MADM) and design (MODM) of alternatives. Except for utility function methods which are used in both MADM and MODM, few applicable methods are available.

Comparative evaluation of MADM methods: Since most methods have not been tested by many real-world problems, we cannot yet discuss in depth the advantages, disadvantages, applicability, computational complexity, difficulty, etc., of each method. A large variety of problems approached by the various methods will provide comparative analysis of the methods. This kind of activity should be carried out and continued for some time to come. The importance of applying MADM methods to solving real-world problems cannot be overemphasized.

read?

VI. BIBLIOGRAPHY

...nographs, and Conference Proceedings:

...ett, J. H., <u>Individual Goals and Organizational Objectives: A Study of Integration Mechanisms,</u> University of Michigan Press, Ann Arbor, Michigan, 1970.

[BM-2] Bechtel, G. G., <u>Multidimensional Preference Scaling</u>, Mouton, Hague, Netherlands, 1976.

[BM-3] Bell, D. E, R. L. Keeney, and H. Raiffa (eds), <u>Conflicting Objectives in Decisions</u>, Wiley, New York, 1977.

[BM-4] Cochrane, J. L. and M. Zeleny (eds.), <u>Multiple Criteria Decision Making,</u> University of South Carolina Press, Columbia, South Carolina, 1973.

[BM-5] Cohon, J. L., <u>Multiobjective Programming and Planning</u>, Academic Press, New York, 1978.

[BM-6] Easton, A., <u>Complex Managerial Decisions Involving Multiple Objectives,</u> Wiley, New York, 1973.

[BM-7] Fishburn, P. C., <u>Utility Theory for Decision Making</u>, Wiley, New York, 1970.

[BM-8] Green, P. E. and F. J. Carmone, <u>Multidimensional Scaling and Related Techniques in Marketing Analysis,</u> Allyn and Bacon, Boston, Massachusetts, 1970.

[BM-9] Green, P. E. and V. R. Rao, <u>Applied Multidimensional Scaling: A Comparison of Approaches and Algorithms,</u> Holt, Rinehart and Winston, New York, 1972.

[BM-10] Green, P. E. and Y. Wind, <u>Multiattribute Decisions in Marketing: A Measurement Approach,</u> Dryden Press, Hinsdale, Illinois, 1973.

[BM-11] Hwang, C. L., A. S. M. Masud, in collaboration with S. R. Paidy and K. Yoon, <u>Multiple Objective Decision Making--Methods and Applications: A State-of-the-Art Survey</u>, Springer-Verlag, Berlin/Heidelberg/New York, 1979.

[BM-12] Johnsen, E., Studies in <u>Multiobjective Decision Models,</u> Monography No. 1 Economic Research Center in Lund, Sweden, 1968.

[BM-13] Keeney, R. L. and H. Raiffa, <u>Decision with Multiple Objectives: Preferences and Value Tradeoffs</u>, Wiley, New York, 1976.

[BM-14] Kruskal, J. B. and M. Wish, <u>Multidimensional Scaling</u>, Sage University Paper Series on Quantitative Applications in the Social Sciences, 07-011, Sage, Beverly Hill and London, 1978.

[BM-15] Leitmann, G. (ed.), <u>Multicriteria Decision Making and Differential Games,</u> Plenum Press, New York, 1976.

[BM-16] Leitmann, G. and A. Marzollo (eds.), Multicriteria Decision Making, Springer-Verlag, Wien/New York, 1975.

[BM-17] Miller, J. R. III, Professional Decision Making: A Procedure for Evaluating Complex Alternatives, Praeger, New York, 1970.

[BM-18] Nijkamp, P. and A. van Delft, Multi-Criteria Analysis and Regional Decision-Making, Martinus Nijhoff Social Sciences Division, Leiden, the Netherlands, 1977.

[BM-19] Shepard, R. M., A. K. Romney, and S. B. Nerlove (eds.), Multidimensional Scaling, Seminar Press, New York and London, 1972.

[BM-20] Starr, M. K. and M. Zeleny (eds.), Multiple Criteria Decision Making, North Holland, New York, 1977.

[BM-21] Thiriez, H. and S. Zionts, (eds.), Multiple Criteria Decision Making: Jouy-en-Josas, France, Springer-Verlag, Berlin/Heidelberg/New York, 1976.

[BM-22] Wendt, D. and C. Vlek (eds.), Utility, Probability, and Human Decision Making, D. Reidel Pub. Co., Boston, 1975.

[BM-23] Zeleny, M. (eds.), Multiple Criteria Decision Making: Kyoto 1975, Springer-Verlag, Berlin/Heidelberg/New York, 1976.

[BM-24] Zionts, S. (ed.), Multiple Criteria Problem Solving: Proceedings, Buffalo, N. Y. (U.S.A.), 1977, Springer-Verlag, Berlin/Heidelberg/New York, 1978.

[BM-17a] Nijkamp, P., Multidimensional Spatial Data and Decision Analysis, Wiley, New York, 1979.

Journal Articles, Technical Reports, and Theses:

1. Adams, E. and R. Fagot, "A Model of Riskless Choice," Behavioral Science, Vol. 3, No. 4, pp. 1-9, 1958.

2. Agnew, N. H., R. A. Agnew, J. Rasmussen, and K. R. Smith, "An Application of Chance Constrained Programming to Portfolio Selection in a Casualty Insurance Firm," Management Science, Vol. 15, No. 10, pp. B512-B520, 1969.

3. Ando, A., "On the Contributions of Herbert A. Simon to Economics," The Scandinavian Journal of Economics, Vol. 81, No. 1, pp. 83-93, 1979.

4. Arrow, K. J., "Utilities, Attitudes, Choices: A Review Note," Econometrica, Vol. 26, No. 1, pp. 1-23, 1958.

5. Aumann, R. J., "Utility Theory without the Complete Axiom," Econometrica, Vol. 30, No. 3, pp. 445-462, 1962.

6. Aumann, R. J., "Utility Theory without the Complete Axiom: A Correction, Econometrica, Vol. 32, No. 1-2, pp. 211-212, 1964.

7. Aumann, R. J., "Subjective Programming," in M. W. Shelly II and G. L. Bryan (eds.), Human Judgment and Optimality, pp. 217-242, Wiley, New York, 1964.

8. Bass, S. M. and H. Kwakernaak, "Rating and Ranking of Multiple-Aspect Alternatives Using Fuzzy Sets," Automatica, Vol. 13, No. 1, pp. 47-58, 1977.

9. Baldwin, J. F. and N. C. F. Guild, "On the Satisfaction of a Fuzzy Relation by a Set of Inputs," International Journal of Man-Machine Studies, Vol. 11, No. 3, pp. 397-404, 1979.

10. Baron, D. P., "Stochastic Programming and Risk Aversion," in Ref. [BM-4], pp. 124-138, 1973.

11. Barron, F. H., "Using Fishburn's Techniques for Analysis of Decision Trees: Some Examples," Decision Sciences, Vol. 4, No. 2, pp. 247-267, 1973.

12. Bauer, V. and M. Wegener, "A Community Information Feedback System with Multiattribute Utilities," in Ref. [BM-3], pp. 323-357, 1977.

13. Baumol, W. J., "An Expected Gain-confidence Limit Criterion for Portfolio Selection," Management Science, Vol. 10, No. 1, pp. 174-182, 1963.

14. Baumol, W. J., "On the Contributions of Herbert A. Simon to Economics," The Scandinavian Journal of Economics, Vol. 81, No. 1, pp. 74-82, 1979.

15. Belk, R. W., "It's the Thought That Counts: A Signed Diagraph Analysis of Gift-Giving," Journal of Consumer Research, Vol. 3, No. 3, pp. 155-162, 1976.

16. Bell, D. E., "A Decision Analysis of Objectives for a Forest Pest Problem," in Ref. [BM-3], pp. 389-424, 1977.

17. Bell, D. E., "Interpolation Independence," in Ref. [BM-24], pp. 1-7, 1978.

18. Bell, D. E., "A Utility Function for Time Streams Having Interperiod Dependencies," Behavioral Science, Vol. 24, No. 3, pp. 208-213, 1979.

19. Bellman, R. E., and L. A. Zadeh, "Decision-Making in a Fuzzy Environment," Management Science, Vol. 17, No. 4, pp. B141-B164, 1970.

20. Benayoun, R., B. Roy, and N. Sussman, "Manual de Reference du Programme Electre," Note de Synthese et Formation, No. 25, Direction Scientifique SEMA, Paris, 1966.

21. Bentler, P. M. and D. G. Weeks, "Restricted Multidimensional Scaling Models," Journal of Mathematical Psychology, Vol. 17, No. 2, pp. 138-151, 1978.

22. Bereanu, B., "Large Group Decision Making with Multiple Criteria," in Ref. [BM-21], pp. 87-102, 1976.

23. Bergstresser, K., A. Charnes, and P. L. Yu, "Generalization of Domination Structures and Nondominated Solutions in Multicriteria Decision Making," Journal of Optimization Theory and Applications, Vol. 18, No. 1, pp. 3-13, 1976.

24. Berhold, M., "Multiple Criteria Decision Making in Consumer Behavior," in Ref. [BM-4], pp. 570-576. 1973.

25. Bernardo, J. J. and J. M. Blin, "A Programming Model of Consumer Choice among Multi-Attributed Brands," Journal of Consumer Research, Vol 4, No. 2, pp. 111-118, 1977.

26. Bettman, J. R., "A Graph Theory Approach to Comparing Consumer Information Processing Models," Management Science, Vol. 18, No. 4 Part II, pp. 114-128, 1971.

27. Bettman, J. R., "Toward a Statistics for Consumer Decision Net Models", Journal of Consumer Research, Vol. 1, No. 2, pp. 71-80, 1974.

28. Bettman, J. R., "A Threshold Model of Attribute Satisfaction Decisions," Journal of Consumer Research, Vol. 1, No. 3, pp. 30-35, 1974.

29. Bettman, J. R., An Information Processing Theory of Consumer Choice, Addison-Wesley, Reading, Mass., 1979, (Chap. 7 Decision Processes: Choice among Alternatives).

30. Bickel, S. H., "Minimum Variance and Optimal Asymptotic Portfolios," Management Science, Vol. 16, No. 3, pp. 221-226, 1969.

31. Bishop, A. B., M. McKee, T. W. Morgan, and R. Narayana, "Multiobjective Planning: Concepts and Methods," Journal of the Water Resources Planning and Management Division, Vol. 102, No. WR2, 1976.

32. Blin, J.M., "The General Concept of Multidimensional Consistency: Some Algebraic Aspects of the Aggregation Problem," in Ref. [BM-4], pp. 164-178, 1973.

33. Blin, J. M., "Fuzzy Relations in Group Decision Theory," Journal of Cybernetics, Vol. 4, No. 2, pp. 17-22, 1974.

34. Blin, J. M., "Fuzzy Sets in Multiple Criteria Decision-Making," in Ref. [BM-20], pp. 129-146, 1977.

35. Blin, J.M.,and J. R. Dodson Jr., "A Multiple Criteria Decision Model for Repeated Choice Siutations," in Ref. [BM-24], pp. 8-22, 1978.

36. Bonoma, T. V., "Business Decision Making," in M. F. Kaplan and S. Schwartz (eds.), Human Judgment and Decision Processes in Applied Settings, pp. 227-254, Academic Press, New York, 1977.

37. Borch, K., "Economic Objectives and Decision Problems", IEEE Trans. on Systems Science and Cybernetics, Vol. SSC-4, No. 3, pp. 266-270, 1968.

38. Borch, K., "Expected Utility Expressed in Terms of Moments," OMEGA, Vol. 1, No. 3, pp. 331-343, 1973.

39. Bowman, E. H. "Consistency and Optimality in Managerial Decision Making," Management Science, Vol. 9, No. 2, pp. 310-321, 1963.

40. Brinskin, L. E., "A Method of Unifying Multiple Objective Functions," Management Science, Vol. 12, No. 10, pp. B406-B416, 1966.

41. Brinskin, L. E., "Establishing a Generalized Multi-Attribute Utility Function," in Ref. [BM-4], pp. 236-245, 1973.

42. Bunn, D. W., "Anchoring Bias in the Assessment of Subjective Probability," Operational Research Quarterly, Vol. 36, No. 2, ii, pp. 449-454, 1975.

43. Burns, T., and L. D. Meeker, "A Mathematical Model of Multi-Dimensional Evaluation, Decision-Making, and Social Interaction," in Ref. [BM-4], pp. 141-163, 1973.

44. Calpine, H. C. and A. Golding, "Some Properties of Pareto-Optimal Choices in Decision Problems," OMEGA, Vol. 4, No. 2, pp. 141-147, 1976.

45. Campbell, V. N. and D. G. Nichols, "Setting Priorities among Objectives," Policy Analysis, Vol. 3, No. 4, pp. 561-578, 1977.

46. Capocelli, R. M. and A. De Luca, "Fuzzy Sets and Decision Theory," Information and Control, Vol. 23, No. 5, pp. 446-473, 1973.

47. Carlson, R. C. and H. H. Therp, "A Multicriteria Approach to Strategic Planning: An Application in Inventory Control," in M. Roubens (ed.), Advances in Operations Research, pp. 75-83, North-Holland, Amsterdam, 1977.

48. Cassidy, R. G., "Urban Housing Selection," Behavioral Science, Vol. 20, No. 4, pp. 241-250, 1975.

49. Chang, S. S. L., "Fuzzy Mathematics, Man, and His Environment," IEEE Trans. on Systems, Man, and Cybernetics, Vol. SMC-2, No. 1, pp. 92-93, 1972.

50. Chang, S. S. L. and L. A. Zadeh, "On Fuzzy Mapping and Control," IEEE Trans. on Systems, Man, and Cybernetics, Vol. SMC-2, No. 1, pp. 30-34, 1972.

51. Charnes, A., W. W. Cooper, and G. Kozmetsky," Measuring, Monitoring and Modeling Quality of Life, "Management Science, Vol. 19, No. 10, pp. 1172-1188, 1973.

52. Chernoff, H., "Rational Selection of Decision Functions," Econometrica, Vol. 22, No. 4, pp. 422-443, 1954.

53. Chipman, J. S., "The Foundations of Utility," Econometrica, Vol. 28, No. 2, pp. 193-224, 1960.

54. Chu, A. T. W., R. E. Kalaba, and K. Spingarn, "A Comparison of Two Methods for Determining the Weights of Belonging to Fuzzy Sets," Journal of Optimization Theory and Applications, Vol. 27, No. 4, pp. 531-538, 1979.

55. Churchman, C. W., "Morality as a Value Criterion," in Ref. [BM-4], pp. 3-8, 1973.

56. Churchman, C. W. and R. L. Ackoff, "An Approximate Measure of Value," Journal of the Operations Research Society of America, Vol. 2, No. 2, pp. 172-187, 1954.

57. Churchman, C. W., R. L. Ackoff, and E. L. Arnoff, Introduction to Operations Research, Wiley, New York, 1957, (Chap. 6 Weighting Objectives).

58. Clarke, D. and B. H. P. Rivett, "A Structural Mapping Approach to Complex Decision-Making," Journal of the Operation Research Society, Vol. 29, No. 2, pp. 113-128, 1978.

59. Clarke, T. E., "Decision-Making in Technologically Based Organizations: A Literature Survey of Present Practice," IEEE Trans. on Engineering Management, Vol. EM-21, No. 1, pp. 9-23, 1974.

60. Cohon, J. L. and D. H. Marks, "A Review and Evaluation of Multi-objective Programming Techniques," Water Resources Research, Vol. 11, No. 2, pp. 208-220, 1975

61. Coombs, C. H., "Inconsistency of Preferences: A Test of Unfolding Theory," in B. M. Fass (ed.), Decision Making, pp. 319-333, Penguin Books, Middlesex, England, 1967.

62. Crawford, A. B., "Impact Analysis Using Differentially Weighted Evaluative Criteria," in Ref. [BM-4], pp. 732-735, 1973.

63. Croley, T. E. II, "Reservoir Operation Through Objectives Trade-Offs," Water Resources Bulletin, Vol. 10, No. 6, pp. 1123-1132, 1974.

64. Dalkey, N. C., "Group Decision Analysis," in Ref. [BM-23], pp. 45-74, 1976.

65. Dalkey, N. C., R. Lewis, and D. Snyder, "Measurement and Analysis of the Quality of Life: with Exploratory Illustration of Applications to Career and Transportation Choice," RAND Memorandum, RM-6228-DOT, 1970.

66. Dasarathy, B. V., "SMART: Similarity Measure Anchored Ranking Technique for the Analysis of Multidimensional Data Analysis," IEEE Trans. on Systems, Man, and Cybernetics, Vol. SMC-6, No. 10, pp. 708-711, 1976.

67. David, L. and L. Duckstein, "Multicriterion Ranking of Alternative Long-range Water Resources Systems," Water Resources Bulletin, Vol. 12, No. 4, pp. 731-754, 1976.

68. Davidson, D. and P. Suppes, "A Finitistic Axiomatization of Subjective Probability and Utility," Econometrica, Vol. 24, No. 3, pp. 264-275, 1956.

69. Davis, O. A., M. H. DeGroot, and M. J. Hinich, "Social Preference Orderings and Majority Rule," Econometrica, Vol. 40, No. 1, pp. 147-157, 1972.

70. Dawes, R. M., "Social Selection Based on Multidimensional Criteria," Journal of Abnormal and Social Psychology, Vol. 68, No. 1, pp. 104-109, 1964.

71. Dawes, R. M., "A Case Study of Graduate Admissions: Applications of Three Principles of Human Decision Making," American Psychologist, Vol. 26, No. 2, pp. 180-188, 1971

72. Dawes, R. M., "Objective Optimization Under Multiple Subjective Functions," in Ref. [BM-4], pp. 9-17, 1973.

73. Dawes, R. M., "Predictive Models as a Guide to Preference," IEEE Trans. on Systems, Man, and Cybernetics, Vol. SMC-7, No. 5, pp. 355-357, 1977.

74. Dawes, R. M. and J. Eagle, "Multivariate Selection of Students in a Racist Society: A Systematically unfair Approach," in Ref. [BM-23], pp. 97-110, 1976.

75. Day, R. H. and S. M. Robinson, "Economic Decisions with L** Utility," in Ref. [BM-4], pp. 84-92, 1973.

76. Dean, R. D. and T. M. Carroll, "Plant Location under Uncertainty," Land Economics, Vol. 53, No. 4, pp. 423-444, 1977.

77. Deutsch, S. D., and J. J. Martin, "An Ordering Algorithm for Analysis of Data Arrays," Operations Research, Vol. 19, No. 6, pp. 1350-1362, 1971.

78. Diamond, P. A., "The Evaluation of Infinite Utility Streams," Econometrica, Vol. 33, No. 1, pp. 170-177, 1965.

79. Dinkel, J. J. and J. E. Erickson, "Multiple Objectives in Environmental Protection Programs," Policy Sciences, Vol. 9, No. 1, pp. 87-96, 1978.

80. Dyer, J. S., W. Farrell, and P. Bradley, "Utility Functions for Test Performance," Management Science, Vol. 20, No. 4, Part I, pp. 507-519, 1973.

81. Dyer, J. S. and R. L. Miles Jr., "Alternative Formulations for a Trajectory Selection Problem: The Mariner Jupiter/Saturn 1977 Project," in Ref. [BM-3], pp. 367-388, 1977.

82. Dyer, J. S. and R. K. Sarin, "Measurable Multiattribute Value Function," Operations Research, Vol. 27, No. 4, pp. 810-822, 1979.

83. Dyer, J. S. and R. K. Sarin, "Cardinal Preference Aggregation Rules for the Case of Certainty," in Ref. [BM-24], pp. 68-86, 1978.

84. Easton, A., "One-of-a-Kind Decisions Involving Weighted Multiple Objectives and Disparate Alternatives," in Ref. [BM-4], pp. 657-667, 1973.

85. Eckenrode, R. T., "Weighting Multiple Criteria," Management Science, Vol. 12, No. 3, pp. 180-192, 1965.

86. Edwards, W., "Use of Multiattribute Utility Measurement for Social Decision Making," in Ref. [BM-3], pp. 247-276, 1977.

87. Edwards, W., "How to Use Multiattribute Utility Measurement for Social Decision Making," IEEE Trans. on Systems, Man, and Cybernetics, Vol. SMC-7, No. 5, pp. 326-340, 1977.

88. Eilon, S., "Goals and Constraints in Decision-making," Operational Research Quarterly, Vol. 23, No. 1, pp. 3-15, 1972.

89. Einhorn, H. J., "The Use of Nonlinear, Noncompensatory Models in Decision Making," Psychological Bulletin, Vol. 73, No. 3, pp. 221-230, 1970

90. Einhorn, H. J. and N. J. Gonedes, "An Exponential Discrepancy Model for Attitude Evaluation," Behavioral Science, Vol. 16, No. 2, pp. 152-157, 1971.

91. Einhorn, H. J. and W. McCoach, "A Simple Multiattribute Utility Procedure for Evaluation," Behavioral Sciences, Vol. 22, No. 4, pp. 270-282, 1977.

92. Einhorn, H. J. and S. Schacht, "Decision Based on Fallible Clinical Judgment," in M. F. Kaplan and S. Schwartz (eds.), Human Judgment and Decision Processes in Applied Settings, pp. 125-144, Academic Press, New York, 1977.

93. Encarnacion, J. Jr., "A Note on Lexicographical Preferences," Econometrica, Vol. 32, No. 1-2, pp. 215-217, 1964.

94. Erlandson, R. F., "System Evaluation Methodologies: Combined Multidimensional Scaling and Ordering Techniques," IEEE Trans. on Systems, Man, and Cybernetics, Vol. SMC-8, No. 6, pp. 421-432, 1978

95. Fandel, G. and J. Wilhelm, "Rational Solution Principles and Information Requirements as Elements of a Theory of Multiple Criteria Decision Making," in Ref. [BM-21], pp. 215-231, 1976.

96. Farguhar, P. H., "A Fractional Hypercube Decomposition Theorem for Multiattribute Utility Function," Operation Research, Vol. 23, No. 5, pp. 941-967, 1975.

97. Farguhar, P. H., "Pyramid and Semicube Decompositions of Multiattribute Utility Functions," Operations Research, Vol. 24, No. 2, pp. 256-271, 1976.

98. Farquhar, P. H., "A Survey of Multiattribute Utility Theory and Applications," in Ref. [BM-20], pp. 59-90, 1977.

99. Farquhar, P. H., "Interdependent Criteria in Utility Analysis," in Ref. [BM-24], pp. 131-180, 1978.

100. Farquhar, P. H., and V. R. Rao, "A Balance Model for Evaluating Subsets of Multiattributed Items," Management Science, Vol. 22, No. 5, pp. 528-539, 1976.

101. Farris, D. R. and A. P. Sage, "Introduction and Survey of Group Decision Making with Applications to Worth Assessment," IEEE Trans. on Systems, Man, and Cybernetics, Vol. SMC-5, No. 3, pp. 346-358, 1975.

102. Fields, D. S., "Cost/Effectiveness Analysis: Its Tasks and Their Interrelation," Operations Research, Vol. 14, No. 3, pp. 515-527, 1966.

103. Findler, N. V., "On the Complexity of Decision Trees, the Quasi-Optimizer, and the Power of Heuristic Rules", Information and Control, Vol. 40, No. 1, pp. 1-19, 1979.

104. Firstman, S. I. and D. S. Stoller, "Establishing Objectives, Measures, and Criteria for Multiphase Complementary Activities," Operations Research, Vol. 14, No. 1, pp. 84-99, 1966.

105. Fischer, G. W., "Experimental Applications of Multi-attribute Utility Models," in Ref. [BM-22], pp. 7-45, 1975.

106. Fischer, G. W., "Convergent Validation of Decomposed Multi-Attribute Utility Assessment Procedures for Risky and Riskless Decisions," Organizational Behavior and Human Performance, Vol. 18, No. 2, pp. 295-315, 1977.

107. Fischer, G. W., "Utility Models for Multiple Objective Decision: Do They Accurately Represent Human Preferences?," Decision Sciences, Vol. 10, No. 3, pp. 451-479, 1979.

108. Fishburn, P. C., "Independence in Utility Theory with Whole Product Sets," Operations Research, Vol. 13, No. 1, pp. 28-45, 1965.

109. Fishburn, P. C., "Analysis of Decisions with Incomplete Knowledge of Probabilities," Operations Research, Vol. 13, No. 2, pp. 217-237, 1965.

110. Fishburn, P. C., "A Note on Recent Development in Additive Utility Theories for Multiple-Factor Situations," Operations Research, Vol. 14, No. 6, pp. 1143-1148, 1966.

111. Fishburn, P. C., "Additive Utilities with Incomplete Product Sets: Applications to Priorities and Assignments," Operations Research, Vol. 15, No. 3, pp. 537-542, 1967.

112. Fishburn, P. C., "Additive Utilities with Finite Sets: Applications in the Management Sciences," Naval Research Logistics Quarterly, Vol. 14, No. 1, pp. 1-13, 1967.

113. Fishburn, P. C., "Methods of Estimating Additive Utilities," Management Science, Vol. 13, No. 7, pp. 435-453, 1967.

114. Fishburn, P. C., "An Abbreviated States of the World Decision Model," IEEE Trans. on Systems Science and Cybernetics, Vol. SSC-4, No. 3, pp. 300-306, 1968.

115. Fishburn, P. C., "Intransitive Indifference in Preference Theory: A Survey," Operations Research, Vol. 18, No. 2, pp. 207-228, 1970.

116. Fishburn, P. C., "The Theorem of the Alternative in Social Choice Theory," Operations Research, Vol. 19, No. 6, pp. 1323-1330, 1971.

117. Fishburn, P. C., "A Mixture-Set Axiomatization of Conditional Subjective Expected Utility," Econometrica, Vol. 41, No. 1, pp. 1-25, 1973.

118. Fishburn, P. C., "Bernoullian Utilities for Multiple-Factor Situations," in Ref. [BM-4], pp. 47-61, 1973.

119. Fishburn, P. C., "von Neumann-Morgenstern Utility Functions on Two Attributes," Operations Research, Vol. 22, No. 1, pp. 35-45, 1974.

120. Fishburn, P. C., "Lexicographic Orders, Utilities and Decision Rules: A Survey," Management Science, Vol. 20, No. 11, pp. 1442-1471, 1974.

121. Fishburn, P. C., "Utility Independence on Subsets of Product Sets", Operations Research, Vol. 24, No. 2, pp. 245-255, 1976.

122. Fishburn, P. C., "Multiattribute Utilities in Expected Utility Theory," in Ref. [BM-3], pp. 172-196, 1977.

123. Fishburn, P. C., "Multicriteria Choice Functions Based on Binary Relations," Operations Research, Vol. 25, No. 6, pp. 989-1012, 1977.

124. Fishburn, P. C., "Stochastic Dominance without Transitive Preferences," Management Science, Vol. 24, No. 12, pp. 1268-1277, 1978.

125. Fishburn, P. C., "Choice Probabilities and Choice Functions," Journal of Mathematical Psychology, Vol. 18, No. 3, pp. 205-219, 1978.

126. Fishburn, P. C., "A Survey of Multiattribute/Multicriterion Evaluation Theories," in Ref. [BM-24], pp. 181-224, 1978.

127. Fishburn, P. C., "Value Theory," in J. J. Moder and S. E. Elmaghraby (eds.), Handbook of Operations Research, Chapter III-3, Van Nostrand Reihold, New York, 1978.

128. Fishburn, P. C. and R. L. Keeney, "Seven Independence Concepts and Continuous Multiattribute Utility Functions," Journal of Mathematical Psychology, Vol. 11, No. 3, pp. 294-372, 1974.

129. Fishburn, P. C. and R. L. Keeney, "Generalized Utility Independence and Some Implications," Operations Research, Vol. 23, No. 5, pp. 928-940, 1975.

130. Fishburn, P. C., A. H. Murphy, and H. H. Isaacs, "Sensitivity of Decisions to Probability Estimation Errors: A Reexamination," Operations Research, Vol. 16, No. 2, pp. 254-267, 1968.

131. Fisk, C. and C. R. Brown, "A Note on the Entropy Formulation of Distribution Models," Operational Research Quarterly, Vol. 26, No. 4, i, pp. 755-758, 1975.

132. Flavell, R., "Simple Decision Problems: A Graphical Solution," OMEGA, Vol. 2, No. 3, pp. 411-414, 1974.

133. Foerster, J. F., "Mode, Choice Decision Process Models: A Comparison of Compensatory and Non-Compensatory Structures," Transportation Research, Vol. 13A, No. 1, pp. 17-28, 1979.

134. Funk, S. G., A. D. Horowitz, R. Lipshitz, and F. W. Young, "The Perceived Structure of American Ethnic Groups: The Use of Multidimensional Scaling in Stereotype Research," Sociometry, Vol. 39, No. 2, pp. 116-130, 1976.

135. Gabriel, K. R., M. Hill, and H. Law-Yone, "A Multivariate Statistical Technique for Regionalization," Journal of Regional Science, Vol. 14, No. 1, pp. 89-106, 1974.

136. Gaines, B. R., "Stochastic and Fuzzy Logics", Electronics Letters, Vol. 11, No. 9, pp. 188-189, 1975.

137. Gaines, B. R. and L. J. Kohout, "The Fuzzy Decade: A Bibliography of Fuzzy Systems and Closely Related Topics," International Journal of Man-Machine Studies, Vol. 9, No. 1, pp. 1-68, 1977.

138. Gardiner, P. C., "Decision Spaces," IEEE Trans. on Systems, Man, and Cybernetics, Vol. SMC-7, No. 5, pp. 340-349, 1977.

139. Gardiner, P. C., and W. Edwards, "Public Values: Multiattribute--Utility Measurement for Social Decision Making," in M. F. Kaplan and S. Schwartz (eds.), Human Judgment and Decision Process, pp. 1-38, Academic Press, New York, 1975.

140. Garner, W. R. and W. J. McGill, "The Relation between Information and Variance Analysis," Psychometrika, Vol. 21, No. 3, pp. 219-228, 1956.

141. Gearing, C. E., W. W. Swart, and T. Var, "Determining the Optimal Investment Policy for the Tourism Sector of A Development Country," Management Science, Vol. 20, No. 4, Part I, pp. 487-497, 1973.

142. Geistfeld, L. V., "Consumer Decision Making: The Technical Efficiency Approach," Journal of Consumer Research, Vol. 4, No. 1, 1977.

143. Giordano, J-L. and J-C. Suquet, "On Multicriteria Decision Making: An Application to a Work-Shop Organization Problem," in M. Roubens (ed.), Advances in Operations Research , pp. 181-192, North Holland, Amsterdam, 1977.

144. Gorman, W. M., "Conditions for Additive Separability," Econometrica, Vol. 36, No. 3-4, pp. 605-609, 1968.

145. Green, P. E., "Multidimensional Scaling and Conjoint Measurement in the Study of Choice Among Multiattribute Alternatives," in Ref. [BM-4], pp. 577-609, 1973.

146. Green, P. E., "On the Design of Choice Experiments Involving Multifactor Alternatives," Journal of Consumer Research, Vol. 1, No. 1, pp. 61-68, 1974.

147. Green, P. E. and F. J. Carmone, "Multidimensional Scaling: An Introduction and Comparison of Nonmetric Unfolding Techniques," Journal of Marketing Research, Vol. 6, No. 3, pp. 330-341, 1969.

148. Green, P. E. and F. J. Carmone, "Evaluation of Multiattribute Alternatives: Additive vs. Configural Utility Measurement," Decision Sciences, Vol. 5, No. 2, pp. 164-181, 1974.

149. Green, P. E. and F. J. Carmone, "Segment Congruence Analysis: A Method for Analyzing Association among Alternative Bases for Market Segment," Journal of Consumer Research, Vol, 3, No. 4, pp. 217-222, 1977.

150. Green, P. E., F. J. Carmone, and D. P. Wachspress, "Consumer Segment via Latent Class Analysis," Journal of Consumer Research, Vol. 3, No. 3, pp. 170-174, 1976.

151. Green, P. E., F. J. Carmone, and Y. Wind, "Subjective Evaluation Models and Conjoint Measurement," Behavioral Science, Vol. 17, No. 3, pp. 288-299, 1972.

152. Green, P. E. and M. T. Devita, "A Complementarity Model of Consumer Utility for Item Collections," Journal of Consumer Research, Vol. 1, No. 3, pp. 56-67, 1974.

153. Green, P. E. and M. T. Devita, "An Interaction Model of Consumer Utility," Journal of Consumer Research, Vol. 2, No. 2, pp. 146-153, 1975.

154. Green, P. E. and V. Srinivasan, "Conjoint Analysis in Consumer Research: Issues and Outlook," Journal of Consumer Research, Vol. 5, No. 2, pp. 103-123, 1978.

155. Green, P. E. and Y. Wind, "New Way to Measure Consumers' Judgments," Harvard Business Review, Vol. 53, No. 4, pp. 107-117, 1975.

156. Green P. E., Y. Wind, and A. K. Jain, "Preference Measurement of Item Collections," Journal of Marketing Research, Vol. 9, No. 4, pp. 371-377, 1972.

157. Gros, J. G., R. Avenhaus, J. Linnerooth, P. O. Pahner, and H. J. Otway, "A Systems Analysis Approach to Nuclear Facility Siting," Behavioral Science, Vol. 21, No. 2, pp. 116-127, 1976.

158. Guigou, J-L, "On French Location Models for Product Units," Regional and Urban Economics, Vol. 1, No. 2, pp. 107-138 and Vol. 1, No. 3, pp. 289-316, 1971.

159. Gulliksen, H. "The Structure of Individual Differences in Optimality Judgments," in M. W. Shelly II and G. L. Bryan (eds.), Human Judgment and Optimality, pp. 72-84, Wiley, New York, 1964.

160. Gum, R. L., T. G. Roefs, and D. B. Kimball, "Quantifying Societal Goals: Development of a Weighting Methodology," Water Resources Research, Vol. 12, No. 4, pp. 612-622, 1976.

161. Gusev, L. A. and I. M. Smirnova, "Fuzzy Sets Theory and Applications (Survey)," Automation and Remote Control, Vol. 34, No. 5, pp. 739-755, 1973.

162. Gustafson, D. H., G. K. Pai, and G. C. Kramer, "A Weighted Aggregate Approach to R & D Project Selection," AIIE Transactions, Vol. 3, No. 1, pp. 22-31, 1971.

163. Hadar, J. and W. R. Russell, "Decision Making with Stochastic Dominance: An Expository Review," OMEGA, Vol. 2, No. 3, pp. 365-377, 1974.

164. Hagerty, M. R., "Model Testing Techniques and Price-Quality Tradeoffs," Journal of Consumer Research, Vol. 5, No. 3, pp. 194-205, 1978.

165. Hammond, K. R., "Externalizing the Parameters of Quasirational Thought," in Ref. [BM-23], pp. 75-96, 1976.

166. Hampton, J. M., "SIMMIDA: A Business Decision-making Game," Operational Research Quarterly, Vol. 27, No. 1, ii, pp. 251-260, 1976.

167. Handa, J., "Risk, Probabilities, and a New Theory of Cardinal Utility," Journal of Political Economy, Vol. 85, No. 1, pp. 97-122, 1977.

168. Hansen, F., "Psychological Theories of Consumer Choice," Journal of Consumer Research, Vol. 3, No. 3, pp. 117-142, 1976.

169. Hansen, P., M. Anciaux-Mundeleer and P. Vincke, "Quasi-Kernels of Outranking Relations," in Ref. [BM-21], pp. 53-63, 1976.

170. Hansen, P. and M. Delattre, "Bicriterion Cluster Analysis as an Exploration Tool," in Ref. [BM-24], pp. 249-273, 1978.

171. Hatry, J. P., "Measuring the Effectiveness of Nondefense Public Programs," Operations Research, Vol. 18, No. 5, pp. 772-784, 1970.

172. Haurie, A. and M. C. Delfour, "Individual and Collective Rationality in a Dynamic Pareto Equilibrium," in Ref. [BM-15], pp. 149-162, 1976.

173. Hauser, J. R. and G. L. Urban, "Assessment of Attribute Importances and Consumer Utility Functions: von Neumann-Morgenstern Theory Applied to Consumer Behavior," Journal of Consumer Research, Vol. 5, No. 4, pp. 251-262, 1979.

174. Hax, A. C. and K. M. Wiig, "The Use of Decision Analysis in Capital Investment Problems," in Ref. [BM-3], pp. 277-297, 1977.

175. Herson, A., "Trade Off Analysis in Environmental Decision Making: An Alternative to Weighted Decision Models," Journal of Environmental Systems, Vol. 7, No. 1, pp. 35-44, 1977.

176. Herstein, I. N. and J. Milnor, "An Axiomatic Approach to Measurable Utility," Econometrica, Vol. 21, No. 2, pp. 291-297, 1953.

177. Herzberger, H. G., "Ordinal Preference and Rational Choice," Econometrica, Vol. 41, No. 2, pp. 188-237, 1973.

178. Heuston, M. C. and G. Ogawa, "Observations on the Theoretical Basis of Cost-Effectiveness," Operations Research, Vol. 14, No. 2, pp. 242-266, 1966.

179. Hildreth, C., "Alternative Conditions for Social Orderings," Econometrica, Vol. 21, No. 1, pp. 81-84, 1953.

180. Hill, M., "Goals-Achievement Matrix for Evaluating Alternative Plans," Journal of the American Institute of Planners, Vol. 34, No. 1, pp. 19-38, 1968.

181. Hill, M. and R. Alterman, "Power Plant Site Evaluation: The Case of the Sharon Plant in Israel," Journal of Environmental Management, Vol. 2, No. 2, pp. 179-196, 1974.

182. Hill, M. and M. Shechter, "Optimal Goal Achievement in the Development of Outdoor Recreation Facilities," in A. G. Wilson (ed.), Urban and Regional Planning, pp. 110-120, Pion, London, 1971.

183. Hill, M. and Y. Tzamir, "Multidimensional Evaluation of Regional Plans Serving Multiple Objectives," Papers of the Regional Science Association, Vol. 29, pp. 139-165, 1972.

184. Hill, M. and E. Werczberger, "Goal Programming and the Goals--Achievement Matrix," International Regional Science Review, Vol. 3, No. 2, pp. 165-181, 1978.

185. Hirsch, G., "The Notion of Characteristic Set and Its Implication for the Analysis and Development of Multicriterion Methods," in Ref. [BM-21], pp. 247-262, 1976.

186. Hirschberg, N. W., "Predicting Performance in Graduate School," in M. F. Kaplan and S. Schwartz (eds.), Human Judgment and Decision Processes in Applied Setting, pp. 95-124, Academic Press, New York, 1977.

187. Hoepfl, R. T. and G. P. Huber, "A Study of Self-explicated Utility Model," Behavioral Science, Vol 15, No. 5, pp. 408-414, 1970.

188. Hogarth, R. M., "Decision Time as a Function of Task Complexity," in Ref. [BM-22], pp. 321-338, 1975.

189. Holloway, C. A., Decision Making under Uncertainty Models and Choices, Prentice-Hall, Englewood Cliffs, N.J., 1979, (Chap. 20 Choices with Multiple Attributes).

190. Holman, E. W., "Completely Nonmetric Multidimensional Scaling," Journal of Mathematical Psychology, Vol. 39, No. 1, pp. 39-51, 1978.

191. Holmes, J. C., "An Ordinal Method of Evaluation," Urban Studies, Vol. 10, No. 1, pp. 179-191, 1973.

192. Hopkins, D. S. P., J-C Larréché, and W. F. Massy, "Multiattribute Reference Functions of University Administrators," in Ref. [BM-23], pp. 287-290, 1976.

193. Houthakker, H. S., "Additive Preferences," Econometrica, Vol. 28, No. 2, pp. 245-257, 1960.

194. Houthakker, H. S., "The Present State of Consumption Theory: A Survey Article," Econometrica, Vol. 29, No. 4, pp. 704-740, 1961.

195. Howard, R. A., "The Foundations of Decision Analysis," IEEE Trans. on Systems Science and Cybernetics, Vol. SSC-4, No. 3, pp. 199-219, 1968.

196. Huber, G. P., "Multi-Attribute Utility Models: A Review of Field and Field-Like Studies," Management Science, Vol. 20, No. 10, pp. 1393-1402, 1974.

197. Huber, G. P., "Methods for Quantifying Subjective Probabilities and Multi-Attribute Utilities," Decision Sciences, Vol. 5, No. 3, pp. 430-458, 1974.

198. Huber, G. P., R. Daneshgar, and D. L. Ford, "An Empirical Comparison of Five Utility Models for Predicting Job Preferences," Organizational Behavior and Human Performance, Vol. 6, No. 3, pp. 267-282, 1971.

199. Huber, G. P., V. K. Sahney, and D. L. Ford, "A Study of Subjective Evaluation Models," Behavioral Science, Vol. 14, No. 6, pp. 483-489, 1969.

200. Huber, J., "Bootstrapping of Data and Decisions," Journal of Consumer Research, Vol. 2, No. 3, pp. 229-234, 1975.

201. Humphreys, P. and A. Humphreys, "An Investigation of Subjective Preference Orderings for Multiattributed Alternatives," in Ref. [BM-22], pp. 119-133, 1975.

202. Hwang, C. L., and K. Yoon, "Principles for Evaluation of Air-Conditioning System," a Progress Report to U. S. Office of Energy from Industrial Engineering, Kansas State University, March 1980.

203. Ito, J., "On Certain Kinds of Orderings," Journal of the Operations Research Society of Japan, Vol. 9, No. 1, pp. 1-15, 1966.

204. Itakura, H., and T. Yamauchi, "A Survey on Multidimensional Scaling Techniques in Multivariate Analysis," Systems and Control, J. of Japan Assoc. of Automatic Control Engineers, (in Japanese), Vol. 29, No. 8, pp. 433-443, 1979.

205. Jacquet-Lagreze, E., "How We Can Use the Notion of Semi-Orders to Build Outranking Relations in Multi-Criteria Decision Making," in Ref. [BM-22], pp. 87-111, 1975.

206. Jacquet-Lagreze, E., "Explicative Models in Multicriteria Preference Analysis," in M. Roubens (ed.), Advances in Operations Research, pp. 213-218, North-Holland, Amsterdam, 1977.

207. Jain, A. K., F. Acito, N. K. Malhotra, and V. Mahajan, "A Comparison of the Internal Validity of Alternative Parameter Estimation Methods in Decompositional Multiattribute Preference Models," Journal of Marketing Research, Vol. 16, No. 3, pp. 312-322, 1979.

208. Jain, R., "Decisionmaking in the Presence of Fuzzy Variables," IEEE Trans. on Systems, Man, and Cybernetics, Vol. SMC-6, No. 10, pp. 698-703, 1976.

209. Jaynes, E. T., "Information Theory and Statistical Mechanics," Physical Review, Vol. 106, No. 4, pp. 620-630, 1957.

210. Jaynes, E. T., "Prior Probabilities," IEEE Trans. on Systems Science and Cybernetics, Vol. SSC-4, No. 3, pp. 227-241, 1968.

211. Johnsen, E., "Experiences in Multiobjective Management Processes," in Ref. [BM-23], pp. 135-152, 1976.

212. Johnson, C. R., "Right-Left Asymmetry in an Eigenvector Ranking Procedure," Journal of Mathematical Psychology, Vol. 19, No. 1, pp. 61-64, 1979.

213. Johnson, E. M. and G. P. Huber, "The Technology of Utility Assessment," IEEE Trans. on Systems, Man, and Cybernetics, Vol. SMC-7, No. 5, pp. 311-325, 1977.

214. Kahne, S., "A Procedure for Optimizing Development Decisions," Automatica, Vol. 11, No. 3, pp. 261-269, 1975.

215. Karmarkar, U. S., "Subjectively Weighted Utility: A Descriptive Extension of the Expected Utility Model," Organizational Behavior and Human Performance, Vol. 21, No. 1, pp. 61-72, 1978.

216. Kassarjian, H. H., "Content Analysis in Consumer Research," Journal of Consumer Research, Vol. 4, No. 1, pp. 8-18, 1977.

217. Kaufmann, A., *Introduction to the Theory of Fuzzy Subsets*, Vol. 1, Academic Press, New York, 1975.

218. Keen, P. G. W., "The Evolving Concept of Optimality," in Ref. [BM-20], pp. 31-58, 1977.

219. Keeney, R. L., "Quasi-Separable Utility Functions," *Naval Research Logistics Quarterly*, Vol. 15, No. 4, pp. 551-566, 1968.

220. Keeney, R. L., "Utility Independence and Preferences for Multiattributed Consequences," *Operations Research*, Vol. 19, No. 4, pp. 875-893, 1971.

221 Keeney, R. L., "An Illustrated Procedure for Assessing Multiattributed Utility Functions," *Sloan Management Review*, Vol. 14, No. 1, pp. 37-50, 1972.

222. Keeney, R. L., "Utility Functions for Multiattributed Consequences," *Management Science*, Vol. 18, No. 5, Part I, pp. 276-287, 1972.

223. Keeney, R. L., "Risk Independence and Multiattributed Utility Functions," *Econometrica*, Vol. 41, No. 1, pp. 27-39, 1973.

224. Keeney, R. L., "Concepts of Independence in Multiattribute Utility Theory," in Ref. [BM-4], pp. 62-71, 1973.

225. Keeney, R. L., "A Decision Analysis with Multiple Objectives: The Mexico City Airport," *The Bell Journal of Economics and Management Science*, Vol. 14, No. 1, 1973.

226. Keeney, R. L., "Multiplicative Utility Functions," *Operations Research*, Vol. 22, No. 1, pp. 22-34, 1974.

227. Keeney, R. L., "Examining Corporate Policy Using Multiattribute Utility Analysis," *Sloan Management Review*, Vol. 17, No. 1, pp.63-76, 1975.

228. Keeney, R. L., "Quantifying Corporate Preferences for Policy Analysis," in Ref. [BM-21], pp. 293-304, 1976.

229. Keeney, R. L., "Decision Analysis," in J. J. Moder and S. E. Elmaghraby (eds.), *Handbook of Operations Research*, Chapter III-4, Van Nostrand Reihold, New York, 1978.

230. Keeney, R. L., "Evaluation of Proposed Storage Sites," *Operations Research*, Vol. 27, No. 1, pp. 49-64, 1979.

231. Keeney, R. L. and G. L. Lilien, "A Utility Model for Product Positioning," in Ref. [BM-24], pp. 321-334, 1978.

232. Keeney, R. L. and K. Nair, "Selecting Nuclear Power Plant Sites in the Pacific Northwest Using Decision Analysis," in Ref. [BM-3], pp. 298-322, 1977.

233. Keeney, R. L. and A. Sicherman, "Assessing and Analyzing Preferences Concerning Multiple Objectives: An Interactive Computer Program," *Behavioral Science*, Vol. 21, No. 3, pp. 173-182, 1976.

234. Keeney, R. L. and E. F. Wood, "An Illustrative Example of the Use of Multiattribute Utility Theory for Water Resources Planning," Water Resources Research, Vol. 13, No. 4, pp. 705-712, 1977.

235. Kirkwood, C. W., "Parametrically Dependent Preferences for Multiattributed Consequences," Operations Research, Vol. 24, No. 1, pp. 92-103, 1976.

236. Kirkwood, C. W., "Social Decision Analysis Using Multiattribute Utility Theory," in Ref. [BM-24], pp. 335-344, 1978.

237. Kirkwood, C. W., "Pareto Optimality and Equity in Social Decision Analysis," IEEE Trans. on Systems, Man, and Cybernetics, Vol. SMC-9, No. 2, pp. 89-91, 1979.

238. Klahr, C. N., "Multiple Objectives in Mathematical Programming," Operations Research, Vol. 6, No. 6, pp. 849-855, 1958.

239. Klahr, D., "Decision Making in a Complex Environment: The Use of Similarity Judgments to Predict Preferences," Management Science, Vol. 15, No. 11, pp. 595-617, 1969.

240. Kleinmuntz, B., "MMPI Decision Rules for the Identification of College Maladjustment: A Digital Computer Approach," Psychology Monographs: General and Applied, Vol. 77, No. 14, pp. 1-22, 1963.

241. Kleiter, G.D., "Dynamic Decision Behavior: Comments on Rapoport's Paper," in Ref. [BM-22], pp. 371-375, 1975.

242. Kabayashi, T., "A Method for Counting the Number of Feasible Subsets of a Partially Ordered Finite Set," Journal of the Operations Research Society of Japan, Vol. 8, No. 4, pp. 155-171, 1966.

243. Kohn, M. C., C. F. Manski, and D. S. Mundel, "An Empirical Investigation of Factors Which Influence College-Going Behavior," Annals of Economic and Social Measurement, Vol. 5, No. 4, pp. 391-419, 1976.

244. Koo, A. Y. C., "Revealed Preferences--A Structural Analysis," Econometrica, Vol. 39, No. 1, pp. 89-97, 1971.

245. Koopmans, T. C., "Stationary Ordinal Utility and Impatience," Econometrica, Vol. 28, No. 2, pp. 287-309, 1960.

246. Koopmans, T. C., "On Flexibility of Future Preference," in M. W. Shelly II and G. L. Bryan (eds.), Human Judgment and Optimality, pp. 243-256, Wiley, New York, 1964.

247. Koopmans, T. C., P. A. Diamond, and R. E. Williamson, "Stationary Utility and Time Perspective," Econometrica, Vol. 32, No. 1-2, pp. 83-100, 1964.

248. Kornbluth, J. S. H., "Duality, Indifference and Sensitivity Analysis in Multiple Objective Linear Programming," Operational Research Quarterly, Vol. 25, No. 4, pp. 599-614, 1974.

249. Kornbluth, J. S. H., "Ranking with Multiple Objectives," in Ref. [BM-24], pp. 345-361, 1978.

250. Kozmetsky, G., C. Wrather, and P. L. Yu, "The Impacts of an Auto Weight Limitation on Energy, Resources, and the Economy," Policy Science, Vol. 9, No. 1, pp. 97-120, 1978.

251. Kruivak, J. A., "Integrating Water Quality and Water Resources Planning," Journal of the Hydraulics Division, Vol. 100, No. NY9, pp. 1257-1262, 1974.

252. Kruskal, J. B., "Multidimensional Scaling by Optimizing Goodness of Fit to a Nonmetric Hypothesis," Psychometrika, Vol. 29, No. 1, pp. 1-27, 1964.

253. Kruskal, J. B., "Nonmetric Multidimensional Scaling: A Numerical Method," Psychometrika, Vol. 29, No. 2, pp. 115-129, 1964.

254. Kruskal, J. B., "Analysis of Factorial Experiments by Estimating Monotone Transformations of the Data," Journal of the Royal Statistical Society, Vol. 27, No. 2, pp. 251-263, 1965.

255. Kruskal, J. B., "Multidimensional Scaling and Other Methods for Discovering Structure," in Enslein, Ralston, and Wilf (eds.), Statistical Methods for Digital Computers, pp. 296-339, Wiley, New York, 1977.

256. Krzysztofowicz, R. and L. Duckstein, "Preference Criterion for Flood Control under Uncertainty," Water Resources Research, Vol. 15, No. 3, pp. 513-520, 1979.

257. Kwakernaak, H., "An Algorithm for Rating Multiple-Aspect Alternatives Using Fuzzy Sets," Memorandum No. 224, Twenty University of Technology, Enschede, Netherlands, 1978; also in Automatica, Vol. 15, No. 5, pp. 615-616, 1979.

258. Larichev, O. I., "A Practical Methodology of Solving Multicriterion Problems with Subjective Criteria," in Ref. [BM-3], pp. 197-208, 1977.

259. LaValle, I. H., "On Admissibility and Bayesness When Risk Attitude But Not the Preference Ranking is Permitted to Vary," in Ref. [BM-4], pp. 72-83, 1973.

260. Legasto, A. A. Jr., "A Multiple-Objective Policy Model: Results of an Application to a Developing Country," Management Science, Vol. 24, No. 5, pp. 498-509, 1978.

261. Leontiff, W., "Introduction to a Theory of the Internal Structure of Functional Relationships," Econometrica, Vol. 15, No. 4, pp. 361-373, 1947.

262. Leung, P., "Sensitivity Analysis of the Effect of Variations in the Form and Parameters of a Multiattribute Utility Model: A Survey," Behavioral Science, Vol. 23, No. 6, pp. 478-485, 1978.

263. Levi, A. M., "Constructive, Extensive Measurement of Preference to Predict Choice Between Sums of Outcomes," Behavioral Science, Vol. 19, No. 5, pp. 326-335, 1974.

264. Lhoas, J., "Multi-criteria Decision Aid Applications to the Selection of the Route for a Pipe-Line," in M. Roubens (ed.), Advances in Operations Research, pp. 265-273, North-Holland, Amsterdam, 1977.

265. Lin, J. G., "Multiple-Objective Optimization: Proper Equality Constraints (PEC) and Maximization of Index Vectors," in Ref. [BM-15], pp. 103-128, 1975.

266. Lingoes, J. C., "The Multivariate Analysis of Qualitative Data," Multivariate Behavioral Research, Vol. 3, No. 1, pp. 61-94, 1968.

267. Litchfield, J. W., "A Research and Development Decision Model Incorporating Utility Theory and Measurement of Social Values," IEEE Trans. on Systems, Man, and Cybernetics, Vol. SMC-6, No. 6, pp. 400-410, 1976.

268. Lucas, H. C. Jr. and J. R. Moore Jr., "A Multiple-Criterion Scoring Approach to Information System Project Selection," INFOR, Vol. 14, No. 1, pp. 1-12, 1976.

269. Luce, R. D., "Semiorders and a Theory of Utility Discrimination," Econometrica, Vol. 24, No. 2, pp. 178-191, 1956.

270. Luce, R. D., "A Probabilistic Theory of Utility," Econometrica, Vol. 26, No. 2, pp. 193-224, 1958.

271. Luce, R. D., "Conjoint Measurement: A Brief Survey," in Ref. [BM-3], pp. 148-171, 1977.

272. Luce, R. D. and D. H. Krantz, "Conditional Expected Utility," Econometrica, Vol. 39, No. 2, pp. 253-271, 1971.

273. MacCrimmon, K. R., "Decision Making Among Multiple-Attribute Alternatives: A Survey and Consolidated Approach," RAND Memorandum, RM-4823-ARPA, 1968.

274. MacCrimmon, K. R., "Improving the System Design and Evaluation Process by the Use of Trade-off Information: An Application to Northeast Corridor Transportation Planning," RAND Memorandum, RM-5877-DOT, 1969.

275. MacCrimmon, K. R., "An Overview of Multiple Objective Decision Making," in Ref. [BM-4], pp. 18-44, 1973.

276. MacCrimmon, K. R., and J. K. Sin, "Making Trade-offs," Decision Sciences, Vol. 5, pp. 680-704, 1974.

277. MacCrimmon, K. R. and M. Toda, "The Experimental Determination of Indifference Curves," The Review of Economic Studies, Vol. 36, No. 4, pp. 433-450, 1969.

278. MacCrimmon, K. R. and D. A. Wehrung, "Trade-off Analysis: The Indifference and Preferred Proportions Approaches," in Ref. [BM-3], pp. 123-147, 1977.

279. MacGill, S. M., "Rectangular Input-Output Tables, Multiplier Analysis and Entropy Maximizing Principles: A New Methodology," Regional Science and Urban Economics, Vol. 8, No. 4., pp. 355-370, 1978.

280. MacKay, D. B. and R. W. Olshavsky, "Cognitive Maps of Retail Locations: An Investigation of Some Basic Issues," Journal of Consumer Research, Vol. 2, No. 3, pp. 197-205, 1975.

281. Majone, G., "The Use of Decision Analysis in the Public Sector," in Ref. [BM-22], pp. 397-408, 1975.

282. Majumdar, T., "Choice and Related Preference," Econometrica, Vol. 24, No. 1, pp. 71-73, 1956.

283. Marschak, J., "Guided Soul-Searching for Multi-Criterion Decisions," in Ref. [BM-23], pp. 1-16, 1976.

284. May, K. O., "A Set of Independent Necessary and Sufficient Conditions for Simple Majority Decision," Econometrica, Vol. 20, No. 4, pp. 680-684, 1952.

285. May, K. O., "Intransitivity, Utility, and the Aggregation of Preference Patterns," Econometrica, Vol. 22, No. 1, pp. 1-13, 1954.

286. McClain, J. O., "Decision Modeling in Case Selection for Medical Utilization Review," Management Science, Vol. 18, No. 2, pp. B706-B717, 1972.

287. McFadden, D., "Quantal Choice Analysis: A Survey," Annals of Economic and Social Measurement, Vol. 5, No. 4, pp. 363-390, 1976.

288. Medanic, J., "Minimax Pareto Optimal Solutions with Application to Linear Quadratic Problems," in Ref. [BM-16], pp. 55-124, 1975.

289. Meyer, R. F., "State-Dependent Time Preference," in Ref. [BM-3], pp. 232-246, 1977.

290. Michalos, A. C., "Rationality between the Maximizers and Satisficers," Policy Sciences, Vol. 4, No. 2, pp. 229-244, 1973.

291. Minas, J. S. and R. L. Ackoff, "Individual and Collective Value Judgments," in M. W. Shelly II and G. L. Bryan (eds.), Human Judgment and Optimality, pp. 351-359, Wiley, New York, 1964.

292. Minnehan, R. F., "Multiple Objectives and Multigroup Decisions Making in Physical Design Situations," in Ref. [BM-4], pp. 506-516, 1973.

293. Mitchell, T. R. and L. R. Beach, "Expectancy Theory, Decision Theory, and Occupational Preference and Choice," in M. F. Kaplan, S. Schwartz (eds.), Human Judgment and Decision Processes in Applied Settings, pp. 203-226, Academic Press, New York, 1977.

294. Moore, L. J., B. W. Taylor III, E. R. Clayton, and S. M. Lee, "Analysis of a Multi-Criteria Project Crashing Model," AIIE Transactions, Vol. 10, No. 2, pp. 163-169, 1978.

295. Morris, W. T., "Value Classification: An Engineering Approach," AIIE Transactions, Vol. 7, No. 4, pp. 356-362, 1975.

296. Morse, J. N., "A Theory of Naive Weights," in Ref. [BM-24], pp. 384-401, 1978.

297. Moscarola, J., "Multicriteria Decision Aid: Two Applications in Education Management," in Ref. [BM-24], pp. 402-423, 1978.

298. Moskowitz, H., "Some Observations on Theories of Collective Decisions," in Ref. [BM-22], pp. 381-396, 1975.

299. Moskowitz, H., and G. P. Wright, Operations Research Techniques for Management, Prentice-Hall, New Jersey, 1979, (Chap. 7 Multiple Objectives and Decision Making).

300. Mottley, C. M. and R. D. Newton, "The Selection of Projects for Industrial Research," Operations Research, Vol. 7, No. 6, pp. 740-751, 1959.

301. Murphy, A. H. and R. L. Winkler, "Subjective Probability Forecasting: Some Real World Experiments," in Ref. [BM-22], pp. 177-198, 1975.

302. Nakayama, H.   et al., "Methodology for Group Decision Support with an Application to Assessment of Residential Environment," IEEE Trans. on Systems, Man, and Cybernetics, Vol. SMC-9, No. 9, pp. 477-485, 1979.

303. Nijkamp, P., "A Multicriteria Analysis for Project Evaluation: Economic-Ecological Evaluation of a Land Reclamation Project," Papers of the Regional Science Association, Vol. 35, pp. 87-111, 1974.

304. Nijkamp, P., "Reflections on Gravity and Entropy Models," Regional Science and Urban Economics, Vol. 5, No. 2, pp. 203-225, 1975.

305. Nijkamp, P., "Stochastic Quantitative and Qualitative Multicriteria Analysis for Environmental Design," Papers of the Regional Science Association, Vol. 39, pp. 175-199, 1977.

306. Nijkamp, P. and J. B. Vos, "A Multicriteria Analysis for Water Resource and Land Use Development," Water Resources Research, Vol. 13, No. 3, pp. 513-518, 1977.

307. North, D. W., "A Tutorial Introduction to Decision Theory," IEEE Trans. on Systems and Cybernetics, Vol. SSC-4, No. 3, pp. 200-210, 1968.

308. Nowakowska, M., "Methodological Problems of Measurement of Fuzzy Concepts in the Social Sciences," Behavioral Science, Vol. 22, No. 2, pp. 107-115, 1977.

309. Nowlan, D. M., "The Use of Criteria Weights in Rank Ordering Techniques of Project Evaluation," Urban Studies, Vol. 12, No. 2, pp. 169-176, 1975.

310. Okuda, T., H. Tanaka, and A. Asai, "A Formulation of Fuzzy Decision Problems with Fuzzy Information Using Probability Measures of Fuzzy Events," Information and Control, Vol. 38, No. 2, pp. 135-147, 1978.

311. Olander, F., "Search Behavior in Non-Simultaneous Choice Situations: Satisficing or Maximizing?," in Ref. [BM-22], pp. 297-320, 1975.

312. Oppenheimer, K. R., "A Proxy Approach to Multi-Attribute Decision Making," Management Science, Vol. 24, No. 6, pp. 675-689, 1978.

313. Ortolano, L., "Water Plan Ranking and the Public Interest," Journal of the Water Resources Planning and Management Division, Vol. 102, No. WRI, pp. 35-48, 1976.

314. Ozernoi, V. M. and M. G. Graft, "Multicriterion Decision Problems," in Ref. [BM-3], pp. 17-39, 1977.

315. Paelinck, J. H. P., "Qualitative Multiple Criteria Analysis, Environmental Protection and Multiregional Development," Papers of the Regional Science Association, Vol. 36, pp. 59-74, 1976.

316. Paine, N. R., "A Useful Approach to the Group Choice Problems," Decision Sciences, Vol. 4, No. 1, pp. 21-30, 1973.

317. Papandreou, A. G., "A Test of a Stochastic Theory of Choice," University of California Publications in Economics, Vol. 16, No. 1, pp. 1-18, 1957.

318. Pardee, F. S. et al., "Measurement and Evaluation of Transportation System Effectiveness," RAND Memorandum, RM-5869-DOT, 1969.

319. Park, C. W., "Seven-Point Scale and a Decision-Maker's Simplifying Choice Strategy: An Operationalized Satisficing-Plus Model," Organizational Behavior and Human Performance, Vol. 21, No. 2, pp. 252-271, 1978.

320. Park, C. W., "A Conflict Resolution Choice Model," Journal of Consumer Research, Vol. 5, No. 2, pp. 124-137, 1978.

321. Parker, B. R. and V. Srinivasan, "A Consumer Preference Approach to the Rural Primary Health-Care Facilities," Operations Research, Vol. 24, No. 5, pp. 991-1025, 1976.

322. Pattanaik, P. K., "Group Choice with Lexicographic Individual Orderings," Behavioral Science, Vol. 18, No. 2, pp. 118-123, 1973.

323. Pearl, J. "A Framework for Processing Value Judgments," IEEE Trans.on Systems, Man and Cybernetics, Vol. SMC-7, No. 5., pp. 349-354, 1977.

324. Pekelman, D. and S. K. Sen, "Mathematical Programming Models for the Determination of Attribute Weights," Management Science, Vol. 20, No. 8, pp. 1217-1229, 1974.

325. Pekelman, D. and S. K. Sen, "Measurement and Estimation of Conjoint Utility Functions," Journal of Consumer Research, Vol. 5, No. 4, pp. 263-271, 1979.

326. Perreault, W. D. Jr. and F. A. Russ, "Comparing Multiattribute Evaluation Process Models," Behavioral Science, Vol. 22, No. 6, pp. 423-431, 1977.

327. Pinkel, B., "On the Decision Matrix and the Judgment Process: A Developmental Decision Example," RAND P-3620, 1969.

328. Pitz, G. F., "A Structural Theory of Uncertain Knowledge," in Ref. [BM-22], pp. 163-176, 1975.

329. Pollak, R. A., "Additive von Neumann-Morgenstern Utility Functions," Econometrica, Vol. 35, No. 3-4, pp. 485-494, 1967.

330. Pratt, J. W., "Risk Aversion in the Small and in the Large," Econometrica, Vol. 32, No. 1-2, pp. 122-136, 1964.

331. Pollak, R. A., "The Risk Independence Axiom," Econometrica, Vol. 41, No. 1, pp. 35-39, 1973.

332. Punj, G. N. and R. Staelin, "The Choice Process for Graduate Business Schools," Journal of Marketing Research, Vol. 15, No. 4, pp. 588-598, 1978.

333. Radner, R., "Mathematical Specification of Goals for Decision Problems," in M. W. Shelly II and G. L. Bryan (eds.), Human Judgment and Optimality, pp. 178-216, Wiley, New York, 1964.

334. Radner, R., "Satisficing," in G. Marchuk (ed.), Optimization Techniques: IFIP Technical Conference, pp. 252-263, Springer-Verlag, New York, 1975.

335. Ragade, R. K., K. W. Hipel, and T. E. Unny, "Nonquantitative Methods in Water Resources Management," Journal of the Water Resources Planning and Management Division, Vol. 102, No. WR2, pp. 297-309, 1976.

336. Raiffa, H., Decision Analysis, Addison-Wesley, Reading, Mass. 1968.

337. Raiffa, H., "Preferences for Multi-attributed Alternatives," RAND Memorandum, RM -5868-DOT/RC, 1969.

338. Rapoport, A., "Research Paradigms for Studying Dynamic Decision Behavior," in Ref. [BM-22], pp. 349-370, 1975.

339. Rapoport, A., "Interpersonal Comparison of Utilities," in Ref. [BM-23], pp. 17-43, 1976.

340. Remus, W. E., "Testing Bowman's Managerial Coefficient Theory Using Competitive Gaming Environment," Management Science, Vol. 24, No. 8, pp. 827-835, 1978.

341. Richard, S. F., "Multivariate Risk Aversion, Utility Independence and Separable Utility Functions," Management Science, Vol. 22, No. 1, pp. 12-21, 1975.

342. Richer, M. K., "Revealed Preference Theory," Econometrica, Vol. 34, No. 3, pp. 635-645, 1966.

343. Rivett, P., "Multidimensional Scaling for Multiobjective Policies," OMEGA, Vol. 5, No. 4, pp. 367-379, 1977.

344. Roy, B., "Problems and Methods with Multiple Objective Functions," Mathematical Programming, Vol. 1, No. 2, pp. 239-266, 1971.

345. Roy, B., "How Outranking Relation Helps Multiple Criteria Decision Making," in Ref. [BM-4], pp. 179-201, 1973.

346. Roy, B., "From Optimization to Multicriteria Decision Aid: Three Main Operational Attitudes," in Ref. [BM-21[pp. 1-34, 1976.

347. Roy, B., "Why Multicriteria Decision Aid May Not Fit in with the Assessment of a Unique Criterion," in Ref. [BM-23], pp. 283-286, 1976.

348. Roy, B., "A Conceptual Framework for a Prescriptive Theory of 'Decision-Aid', in Ref. [BM-20], pp. 179-210, 1977.

349. Roy, B., "Partial Preference Analysis and Decision-Aids: The Fuzzy Outranking Relation Concept," in Ref. [BM-3], pp. 40-75, 1977.

350. Saaty, T. L., "A Scaling Method for Priorities in Hierarchical Structures," Journal of Mathematical Psychology, Vol. 15, No. 3, pp. 234-281, 1977.

351. Saaty, T. L., "Exploring the Interface Between Hierarchies, Multiple Objectives and Fuzzy Sets," Fuzzy Sets and Systems, Vol. 1, No. 1, pp. 57-68, 1978.

352. Saaty, T. L. and M. W. Khouja, "A Measure of World Influence," Journal of Peace Science, Vol. 2, No. 1, pp. 31-48, 1976.

353. Sarin, R. K., "Interactive Evaluation and Bound Procedure for Selecting Multi Attributed Alternatives," in Ref. [BM-20], pp. 211-224, 1977.

354. Sarin, R. K., "Screening of Multiattribute Alternatives," OMEGA, Vol. 5, No. 4, pp. 481-487, 1977.

355. Sarin, R. K., "An Interactive Procedure for Subset Selection with Ordinal Preferences," IEEE Trans. on Systems, Man and Cybernetics, Vol. SMC-8, No. 10, pp. 760-763.

356. Sayeki, Y. and K. H. Vesper, "Allocation of Importance in a Hierarchial Goal Structure," Management Science, Vol. 19, No. 6, pp. 667-675, 1973.

357. Selvidge, J., "A Three-Step Procedure for Assigning Probabilities to Rare Events," in Ref. [BM-22], pp. 199-218, 1975.

358. Schlager, K., "The Rank-Based Expected Value Method of Plan Evaluation," Highway Research Record, No. 238, pp. 153-158, 1968.

359. Schmitendorf, W. E. and G. Moriarty, "A Sufficiency Condition for Coalitive Pareto-Optimal Solutions," in Ref. [BM-15], pp. 163-173, 1976.

360. Schwartz, L. E., "Uncertainty Reduction Over Time in the Theory of Multiattributed Utility," in Ref. [BM-4], pp. 108-123, 1973.

361. Schwartz, S. L., I. Vertinsky, and W. T. Ziemba, "R & D Project Selection Behavior: Study Designs and Some Pilot Results," in Ref. [BM-21], pp. 136-146, 1976.

362. Schwartz, S. L., I. Vertinsky, W. T. Ziemba, and M. Bernstein, "Some Behavioural Aspects of Information Use in Decision Making: A Study of Clinical Judgments," in Ref. [BM-21], pp. 378-391, 1976.

363. Seiler, K., "A Cost Effectiveness Comparison Involving A Tradeoff Performance, Cost, and Obtainability," Operations Research, Vol. 14, No. 3, pp. 528-531, 1966.

364. Sengupta, S. S., M. L. Podrebarac, and T. D. H. Fernando, "Probabilities of Optima in Multi-Objective Linear Programmes," in Ref. [BM-4], pp. 217-235, 1973.

365. Shannon, C. E. and W. Weaver, The Mathematical Theory of Communication, The University of Illinois Press, Urbana, Ill., 1947.

366. Shepard, R. N., "The Analysis of Proximities," Part I and II, Psychometrica, Vol. 27, pp. 125-139 and 219-246, 1962.

367. Shepard, R. N., "On Subjectively Optimum Selection Among Multiattribute Alternatives," in M. W. Shelly II and G. L. Bryan (eds.), Human Judgments and Optimality, pp. 257-281, Wiley, New York, 1964.

368. Sheridan, T. D. and A. Sicherman, "Estimation of Group's Multiattribute Utility Function in Real Time by Anonymous Voting," IEEE Trans. on Systems, Man, and Cybernetics, Vol. SMC-7, No. 5, pp. 392-394.

369. Shimura, M., "Fuzzy Sets Concepts in Rank-Ordering Objects," Journal of Mathematical Analysis and Applications, Vol. 43, No. 3, pp. 717-733, 1973.

370. Shocker, A. D. and V. Srinivasan, "A Consumer-Based Methodology for the Identification of New Product Ideas," Management Science, Vol. 20, No. 6, pp. 921-937, 1974.

371. Shocker, A. D. and V. Srinivasan, "Multiattribute Approaches for Product Concept Evaluation and Generation: A Critical Review," Journal of Marketing Research, Vol. 16, No. 2, pp. 159-180, 1979.

372. Siegel, S., "A Method for Obtaining an Ordered Metric Scale," Psychometrica, Vol. 21, No. 2, pp. 207-216, 1956.

373. Simon, H. A., "A Behavioral Model of Rational Choice," Quarterly Journal of Economics, Vol. 69, No. 1, pp. 99-114, 1955.

374. Sinden, J. A. and J. B. Wyckoff, "Indifference Mapping: An Empirical Methodology for Economic Evaluation of the Environment," Regional Science and Urban Economics, Vol. 6, No. 1, pp. 81-103, 1976.

375. Sizer, P. W., "A Behavioral Model of Company Development," in Ref. [BM-21], pp. 392-401, 1976.

376. Smith, L. H., R. W. Lawless, and B. Shenoy, "Evaluation Multiple Criteria-Models for Two-Criteria Situations," Decision Sciences, Vol. 5, No. 4, pp. 587-596, 1974.

377. Smith, R. D. and P. S. Greenlaw, "Simulation of a Psychological Decision Process in Personnel Selection," Management Science, Vol. 13, No. 8, pp. B409-B419, 1967.

378. Spetzler, C. S., "The Development of a Corporate Risk Policy for Capital Investment Decisions," IEEE Trans. on Systems Science and Cybernetics, Vol. SSC-4, No. 3, pp. 279-300, 1968.

379. Srinivasan, V., "Network Models for Estimating Brand-Specific Effects in Multi-Attribute Marketing Models," Management Science, Vol. 25, No. 1, pp. 11-21, 1979.

380. Srinivasan, V. and A. D. Shocker, "Linear Programming Techniques for Multidimensional Analysis of Preference," Psychometrika, Vol. 38, No. 3, pp. 337-369, 1973.

381. Srinivasan, V. and A. D. Shocker, "Estimating the Weights for Multiple Attributes in a Composite Criterion Using Pairwise Judgments," Psychometrika, Vol. 38, No. 4, pp. 473-493, 1973.

382. Stadler, W., "Sufficient Conditions for Preference Optimality," in Ref. [BM-15], pp. 129-148, 1976.

383. Stadler, W., "Preference Optimality (On Optimality Concept in Multicriteria Problems)," in W. Oettli and K. Ritter (eds.), Optimization and Operations Research, pp. 297-306, Springer-Verlag, New York, 1976.

384. Starr, M. K. and L. H. Greenwood, "Normative Generation of Alternatives with Multiple Criteria Evaluation," in Ref. [BM-20], pp. 111-128, 1977.

385. Starr, M. K. and M. Zeleny, "MCDM-State and Future of the Arts," in Ref. [BM-20], pp. 5-29, 1977.

386. Stevens, S. S., "Measurement, Psychophysics, and Utility," in C. W. Churchman and P. Ratoosh (eds.), Measurement--Definitions and Theories, pp. 18-63, Wiley, 1959.

387. Stigum, B. P., "Finite State Space and Expected Utility Maximization," Econometrica, Vol. 40, No. 2, pp. 253-259, 1972.

388. Stimson, D. H., "Utility Measurement in Public Health Decision Making," Management Science, Vol. 16, No. 2, pp. B17-B30, 1969.

389. Strotz, R. H., "The Empirical Implications of A Utility Tree," Econometrica, Vol. 25, No. 2, pp. 269-280, 1957.

390. Strum, J., "Eigenvalues for the Decision Sciences," Decision Sciences, Vol. 4, No. 4, pp. 533-548, 1973.

391. Swalm, R. O., "Utility Theory-Insights into Risk Taking," Harvard Business Review, Vol. 44, No. 6, pp. 123-136, 1966.

392. Swinth, R. L., "Organizational Joint Problem-Solving," Management Science, Vol. 18, No. 2, pp. B68-B79, 1971.

393. Takahara, Y. and K. Kijima, "Characterization of the Satisfactory Decision Principle," Journal of the Operations Research Society of Japan, Vol. 21, No. 3, pp. 347-369, 1978.

394. Takane, Y. and F. W. Young, "Nonmetric Individual Differences Multidimensional Scaling: An Alternative Least Squares Method with Optimal Scaling Features," Psychometrika, Vol. 42, No. 1, pp. 7-67, 1977.

395. Tapiero, C. S., "The Theory of Graphs in Behavioral Science," Decision Sciences, Vol. 3, No. 1, pp. 57-81, 1972.

396. Teger, S., "Presidential Stragety for the Appointment of Supreme Court Justices," Public Choice, Vol. 31, pp. 1-22, 1977.

397. Tell, B., "A Comparative Study of Four Multiple-Criteria Methods," in Ref. [BM-21], pp. 183-197, 1976.

398. Tell, B., "The Effect of Uncertainty on the Selection of a Multiple-Criteria Utility Model," in M. Roubens (ed.), Advances in Operations Research, pp. 497-504, North-Holland, Amsterdam, 1977.

399. Tell, B., "An Approach to Solve Multi-Person Multi-Criteria Decision-Making Problems," in Ref. [BM-24], pp. 482-493, 1978.

400. Terry, H., "Comparative Evaluation of Performance Using Multiple Criteria," Management Science, Vol. 9, No. 3, pp. 431-442, 1963.

401. The Technical Committee of the Water Resources Centers of the Thirteen Western States, "Water Resources Planning and Social Goals: Conceptualization Toward a New Methodology," Utah Water Research Laboratory Publication PRWG-94-1, Utah State University, Logan, Utah, 1971.

402. Thiriez, H. and D. Houri, "Multi-Person Multi-Criteria Decision-Making: A Sample Approach," in Ref. [BM-21], pp. 103-119, 1976.

403. Tilley, P. and S. Eilon, "Stochastic Dominance for Ranking Ventures," OMEGA, Vol. 3, No. 2, pp. 177-184, 1975.

404. Togsverd, T., "Multi-Level Planning in the Public Sector," in Ref. [BM-21], pp. 201-214, 1976.

405. Torgerson, W. S., Theory and Methods of Scaling, Wiley, New York, 1958.

406. Tomlinson, J. W. C. and I. Vertinsky, "Selecting a Strategy for Joint Venture in Fisheries: A First Approximation," in Ref. [BM-21], pp. 351-363, 1976.

407. Turban, E. and M. L. Metersky, "Utility Theory Applied to Multivariable System Effectiveness Evaluation," Management Science, Vol. 17, No. 12, pp. B817-B828, 1971.

408. Tversky, A., "Additivity, Utility and Subjective Probability," Journal of Mathematical Psychology, Vol. 4, No. 2, pp. 175-202, 1967.

409. Tversky, A., "Intransitivity of Preferences," Psychological Review, Vol. 76, No. 1, pp. 31-48, 1969.

410. Tversky, A., "Elimination by Aspects: A Probabilistic Theory of Choice," Michigan Mathematical Psychology Program MMPP 71-12, The University of Michigan, Ann Arbor, Michigan, 1971.

411. Tversky, A., "On the Elicitation of Preferences: Descriptive and Prescriptive Considerations," in Ref. [BM-3], pp. 209-222, 1977.

412. Tversky, A. and D. Kahneman, "Judgment under Uncertainty: Heuristics and Biases," in Ref. [BM-22], pp. 141-162, 1975.

413. van Delft, A. and P. Nijkamp, "A Multi-objective Decision Model for Regional Development, Environmental Quality Control and Industrial Land Use," Papers of the Regional Science Association, Vol. 36, pp. 35-57, 1976.

414. van Praag, B. M. S., "Utility, Welfare, and Probability: An Unorthodox Economist's View," in Ref. [BM-22], pp. 279-296, 1975.

415. Vargas, G., M. Hottenstein, and S. Aggarwal, "Using Utility Functions for Aggregate Scheduling in Health Maintenance Organizations (HMO's)," AIIE Transactions, Vol. 11, No. 4, pp. 327-335, 1979.

416. Vedder, J. N., "Planning Problems with Multidimensional Consequences," Journal of the American Institute of Planners, Vol. 36, No. 2, pp. 112-119, 1970.

417. Vedder, J. N., "Multiattribute Decision Making Under Uncertainty Using Bounded Intervals," in Ref. [BM-4], pp. 93-107, 1973.

418. Vickson, R. G., "Stochastic Dominance Tests for Decreasing Absolute Risk Aversion I: Discrete Random Variables," Management Science, Vol. 21, No. 12, pp. 1438-1446, 1975.

419. Vickson, R. G., "Stochastic Dominance Tests for Decreasing Absolute Risk Aversion II: General Random Variables," Management Science, Vol. 23, No. 5, pp. 478-489, 1977.

420. Vincke, P., "A New Approach to Multiple Criteria Decision-Making," in Ref. [BM-21], pp. 341-350, 1976.

421. von Neumann, J. and O. Morgenstern, Theory of Games and Economic Behavior, Princeton University Press, Princeton, N.J., 1944.

422. von Winterfeldt, D., "Multi-Criteria Decision Making: Comment on Jacquet-Lagreze's Paper," in Ref. [BM-22], pp. 113-117, 1975.

423. von Winterfeldt and G. W. Fischer, "Multiattribute Utility Theory: Models and Assessment Procedures," in Ref. [BM-22], pp. 47-85, 1975.

424. Watanabe, S., "A Generalized Fuzzy-Set Theory," IEEE Trans.on Systems, Man, and Cybernetics, Vol. SMC-8, No. 10, pp. 756-760, 1978.

425. Watson, S. R., J. J. Weiss, and M. L. Donnell, "Fuzzy Decision Analysis," IEEE Trans. on Systems, Man, and Cybernetics, Vol. SMC-9, No. 1, pp. 1-9, 1979.

426. Wehrung, D. A., J. F. Bassler, K. R. MacCrimmon, and W. T. Stanbury, "Multiple Criteria Dominance Models: An Empirical Study of Investment Preferences," in Ref. [BM-24], pp. 494-508, 1978.

427. Weisberg, H. F. and J. G. Rusk, "Dimensions of Candidate Evaluation," The American Political Science Review, Vol. 64, No. 10, pp. 1167-1185, 1970.

428. Westwood, D., T. Lunn, and D. Beazley, "The Trade-off Model and Its Extensions," Journal of the Market Research Society, Vol. 16, No. 3, pp. 227-241, 1974.

429. White, D. J., "Entropy and Decision," Operational Research Quarterly, Vol. 26, No. 1, pp. 15-23, 1975.

430. Wiedey, G. and H. J. Zimmermann, "Media Selection and Fuzzy Linear Programming," Journal of the Operational Research Society, Vol. 29, No. 11, pp. 1071-1084, 1978.

431. Willis, R. E., "A Simulation of Multiple Selection Using Nominal Group Procedures," Management Science, Vol. 25, No. 2, pp. 171-181, 1979.

432. Wilson, A. G., "The Use of the Concept of Entropy in System Modeling," Operational Research Quarterly, Vol. 21, No. 2, pp. 247-265, 1970.

433. Wilson, R. B., "Decision Analysis in a Corporation," IEEE Trans. on Systems Science and Cybernetics, Vol. SSC-4, No. 3, pp. 220-226, 1968.

434. Winterfeldt, D. V. and G. W. Fischer, "Multiattribute Utility Theory: Models and Assessment Procedures," in Ref. [BM-22], pp. 47-85, 1975.

435. Wright, P. and F. Barbour, "Phased Decision Strategies: Sequels to an Initial Screening," in Ref. [BM-20], pp. 91-109, 1977.

436. Yager, R. R., "Multiple Objective Decision-Making Using Fuzzy Sets," International Journal of Man-Machine Studies, Vol. 9, No. 4, pp. 375-382, 1977.

437. Yager, R. R., "A Measurement-Informational Discussion of Fuzzy Union and Intersection," International Journal of Man-Machine Studies, Vol. 11, No. 2, pp. 189-200, 1979.

438. Yager, R. R., "Possibility Decision Making," IEEE Trans. on Systems, Man, and Cybernetics, Vol. SMC-9, No. 7, pp. 388-392, 1979.

439. Yager, R. R., "An Eigenvalue Method of Obtaining Subjective Probabilities," Behavioral Science, Vol. 24, No. 6, 1979.

440. Yager, R. R. and D. Basson, "Decision Making with Fuzzy Sets," Decision Sciences, Vol. 6, No. 3, pp. 590-600, 1975.

441. Young, D. R., "Choosing Among Alternative Complex Systems When Input Characteristics are Uncertain," IEEE Trans. on Systems, and Cybernetics, Vol. SMC-1, No. 1, pp. 77-82, 1971.

442. Yu, P. L., "A Class of Solutions for Group Decision Problems," *Management Science*, Vol. 19, No. 8, pp. 936-946, 1973.

443. Yu, P. L., "Introduction to Domination Structures in Multicriteria Decision Problems," in Ref. [BM-4], pp. 249-261, 1973.

444. Yu, P. L., "Cone Convexity, Cone Extreme Points, and Non-dominated Solutions in Decision Problems with Multiobjectives," *Journal of Optimization Theory and Applications*, Vol. 14, No. 3, pp. 319-376, 1974.

445. Yu, P. L., "Domination Structures and Nondominated Solutions," in Ref. [BM-16], pp. 227-280, 1975.

446. Yu, P. L., "Decision Dynamics with an Application to Persuasion and Negotiation," in Ref. [BM-20], pp. 159-178, 1977.

447. Yu, P. L., "Second-Order Game Problems: Decision Dynamics in Gaming Phenomena," *Journal of Optimization Theory and Applications*, Vol. 27, No. 1, pp. 147-166, 1979.

448. Yu, P. L. and G. Leitmann, "Compromise Solutions, Domination Structures, and Salukvadze's Solution," *Journal of Optimization Theory and Applications*, Vol. 13, No. 3, pp. 362-378, 1974.

449. Yu, P. L. and G. Leitmann, "Nondominated Decisions and Cone Convexity in Dynamic Multicriteria Decision Problems," in Ref. [BM-15], pp. 61-72, 1976.

450. Yu, P. L. and G. Leitmann, "Confidence Structures in Decision Making," *Journal of Optimization Theory and Applications*, Vol. 22, No. 2, pp. 265-285, 1977.

451. Zachow, E. W., "Positive-Difference Structures and Bilinear Utility Functions," *Journal of Mathematical Psychology*, Vol. 17, No. 2, pp. 152-164, 1978.

452. Zadeh, L. A., "Fuzzy Sets," *Information and Control*, Vol. 8, No. 3, pp. 338-353, 1965.

453. Zadeh, L. A., "Fuzzy Algorithms," *Information and Control*, Vol. 12, No. 2, pp. 94-102, 1968.

454. Zadeh, L. A., "Quantitative Fuzzy Semantics," *Information Sciences*, Vol. 3, No. 2, pp. 159-176, 1971.

455. Zadeh, L. A., "Similarity Relations and Fuzzy Orderings," *Information Sciences*, Vol. 3, No. 2, pp. 177-200, 1971.

456. Zadeh, L. A., "Outline of a New Approach to the Analysis of Complex Systems and Decision Processes," IEEE Trans. on Systems, Man, and Cybernetics Vol. SMC-3, No. 1, pp. 28-44, 1973.

457. Zadeh, L. A., "Fuzzy Sets," in A. G. Holzman (ed.), *Operations Research Support Methodology*, pp. 569-606, Marcel Dekker, New York, 1979.

458. Zeleny, M., "A Selected Bibliography of Works Related to Multiple Criteria Decision Making," in Ref. [BM-4], pp. 779-796, 1973.

459. Zeleny, M., Linear Multiobjective Programming, Springer-Verlag, Berlin/ Heidelberg/New York, 1974.

460. Zeleny, M., "A Concept of Compromise Solutions and the Method of the Displaced Ideal," Computers and Operations Research, Vol. 1, No. 4, pp. 479-496, 1974.

461. Zeleny, M., "The Theory of the Displaced Ideal," in Ref. [BM-23], pp. 153-206, 1976.

462. Zeleny, M., "MCDM Bibliography - 1975," in Ref. [BM-23], pp. 291-321, 1976.

463. Zeleny, M., "The Attribute-Dynamic Attitude Model (ADAM)," Management Science, Vol. 23, No. 1, pp. 12-26, 1976.

464. Zeleny, M., "Adaptive Displacement of Preferences in Decision Making," in Ref. [BM-20], pp. 147-158, 1977.

465. Zeleny, M., "Multidimensional Measure of Risk: Prospect Rating Vector," in Ref. [BM-24], pp. 529-548, 1978.

466. Zeleny, M. and J. L. Cochrane, "A Priori and a Posteriori Goals in Macroeconomic Policy Making," in Ref. [BM-4], pp. 373-391, 1973.

467. Zionts, S., "Multiple Criteria Decision Making for Discrete Alternatives with Ordinal Criteria," Working Paper No. 299, School of Management, State University of New York at Buffalo, 1977.

468. Zionts, S., "MCDM--If Not a Roman Numeral, Then What?," Working Paper No. 358, School of Management, State University of New York at Buffalo, 1978.

469. Zionts, S., "Methods for Solving Management Problems Involving Multiple Objectives," Working Paper No. 400, School of Management, State University of New York at Buffalo, 1979.

Supplement

41a. Brown, R. V., A. S. Kahn, and C. Peterson, Decision Analysis for the Manager, Holt, Rinehart and Winston, New York, 1974.

41b. Browning, J. E., How to Select a Business Site, McGraw-Hill, New York, 1980.

51a. Charnetski, J. R., "Multiple Criteria Decisionmaking with Partial Information: A Site Selection Problem," in M. Chatterji (ed.), Space Location + Regional Development, pp. 51-62, Pion, London, 1976.

69a. Davos, C. A., "A Priority-tradeoffs-scanning Approach to Evaluation in Environmental Management," Journal of Environmental Management, Vol. 5, No. 3, pp. 259-273, 1977.

69b. Davos, C. A., C. J. Smith, and N. W. Neinberg, "An Application of the Priority-Tradeoff-Scanning Approach: Electric Power Plant Siting and Technology Evaluation," Journal of Environmental Management, Vol. 8, No. 2, pp. 105-125, 1979.

79a. Dushnik, B. and E. W. Miller, "Partially Ordered Sets," American Journal of Mathematics, Vol. 63, No. 3, pp. 600-610, 1941.

90a. Einhorn, H. J. and R. M. Hogarth,"Unit Weighting Schemes for Decision Making," Organizational Behavior and Human Performance, Vol. 13, No. 2, pp. 171-192, 1975.

92a. Ellis, H. M. and R. L. Keeney, "A Rational Approach to Government Decisions Concerning Air Pollution," in A. W. Drake, R. L. Keeney, and P. M. Morse (eds.), Analysis of Public Systems, pp. 376-400, The MIT Press, Cambridge, Mass., 1972.

116a. Fishburn, P. C., "A Comparative Analysis of Group Decision Methods," Behavioral Science, Vol. 16, No. 6, pp. 538-544, 1971.

156a. Grochow, J. M., "On User Supplied Evaluations of Time Shared Computer Systems," IEEE Trans. on Systems, Man, and Cybernetics, Vol. SMC-3, No. 2, pp. 204-206, 1973.

163a. Haefele, E. T., Representative Government and Environmental Management, The Johns Hopkins University Press, Baltimore, 1973, (Chpater 2).

178a. Hey, J. D., Uncertainty in Microeconomics, Martin Robertson, Oxford, 1979.

214a. Kahneman, D. and A. Tversky, "Subjective Probability: A Judgment of Representativeness," Cognitive Psychology, Vol. 3, No. 3, pp. 430-454, 1972.

239a. Klee, A. J., "The Role of Decision Models in the Evaluation of Competing Environmental Health Alternatives," Management Science, Vol. 18, No. 2, pp. B52-B67, 1971.

302a.   Navarrete, N. Jr., M. Fukushima, and H. Mine, "A New Ranking Method Based on Relative Position Estimate and Its Extensions," IEEE Trans. on Systems, Man, and Cybernetics, Vol. SMC-9, No. 11, pp. 681-689, 1979.

326a.   Peterson, C. R., and L. R. Beach, "Man as an Intuitive Statistician," Psychological Bulletin, Vol. 68, pp. 29-46, 1967.

350a.   Saaty, T. L., "The Sudan Transportation Study," Interfaces, Vol. 8, No. 1, Part 2, pp. 37-57, 1977.

355a.   Savage, L. J., The Foundations of Statistics, Wiley, New York, 1954.

358a.   Schlaiffer, R., Probability and Statistics for Business Decisions, McGraw Hill, New York, 1959.

365a.   Shapley, L. S. and M. Shubik, "A Method for Evaluating the Distribution of Power in a Committee System," The American Political Science Review, Vol. 48, No. 3, pp. 787-792, 1954.

375a.   Slovic, P., and S. Lichtenstein, "Comparison of Bayesian and Regression Approaches to the Study of Information Processing Judgments," Organizational Behavior and Human Performance, Vol. 6, No. 6., pp. 649-744, 1971.

402a.   Thrall, R. M., C. H. Coombs, and R. L. Davis, Decision Process, Wiley, New York, 1954.

410a.   Tversky, A., "Elimination by Aspects: A Theory of Choice," Psychological Review, Vol. 79, No. 4, pp. 281-299, 1972.

410b.   Tversky, A., "Choice by Elimination," Journal of Mathematical Psychology, Vol. 9, No. 4, pp. 341-367, 1972.

440a.   Yoon, K., "Systems Selection by Multiple Attribute Decision Making," Ph.D. Dissertation, Kansas State University, Manhattan, Kansas, 1980.

440b.   Yoon, K. and C. L. Hwang, "TOPSIS (Technique for Order Preference by Similarity to Ideal Solution--A Multiple Attribute Decision Making," a paper to be published, 1980.

Vol. 83: NTG/GI-Gesellschaft für Informatik, Nachrichtentechnische Gesellschaft. Fachtagung „Cognitive Verfahren und Systeme", Hamburg, 11.–13. April 1973. Herausgegeben im Auftrag der NTG/GI von Th. Einsele, W. Giloi und H.-H. Nagel. VIII, 373 Seiten. 1973.

Vol. 84: A. V. Balakrishnan, Stochastic Differential Systems I. Filtering and Control. A Function Space Approach. V, 252 pages. 1973.

Vol. 85: T. Page, Economics of Involuntary Transfers: A Unified Approach to Pollution and Congestion Externalities. XI, 159 pages. 1973.

Vol. 86: Symposium on the Theory of Scheduling and its Applications. Edited by S. E. Elmaghraby. VIII, 437 pages. 1973.

Vol. 87: G. F. Newell, Approximate Stochastic Behavior of n-Server Service Systems with Large n. VII, 118 pages. 1973.

Vol. 88: H. Steckhan, Güterströme in Netzen. VII, 134 Seiten. 1973.

Vol. 89: J. P. Wallace and A. Sherret, Estimation of Product. Attributes and Their Importances. V, 94 pages. 1973.

Vol. 90: J.-F. Richard, Posterior and Predictive Densities for Simultaneous Equation Models. VI, 226 pages. 1973.

Vol. 91: Th. Marschak and R. Selten, General Equilibrium with Price-Making Firms. XI, 246 pages. 1974.

Vol. 92: E. Dierker, Topological Methods in Walrasian Economics. IV, 130 pages. 1974.

Vol. 93: 4th IFAC/IFIP International Conference on Digital Computer Applications to Process Control, Part I. Zürich/Switzerland, March 19–22, 1974. Edited by M. Mansour and W. Schaufelberger. XVIII, 544 pages. 1974.

Vol. 94: 4th IFAC/IFIP International Conference on Digital Computer Applications to Process Control, Part II. Zürich/Switzerland, March 19–22, 1974. Edited by M. Mansour and W. Schaufelberger. XVIII, 546 pages. 1974.

Vol. 95: M. Zeleny, Linear Multiobjective Programming. X, 220 pages. 1974.

Vol. 96: O. Moeschlin, Zur Theorie von Neumannscher Wachstumsmodelle. XI, 115 Seiten. 1974.

Vol. 97: G. Schmidt, Über die Stabilität des einfachen Bedienungskanals. VII, 147 Seiten. 1974.

Vol. 98: Mathematical Methods in Queueing Theory. Proceedings 1973. Edited by A. B. Clarke. VII, 374 pages. 1974.

Vol. 99: Production Theory. Edited by W. Eichhorn, R. Henn, O. Opitz, and R. W. Shephard. VIII, 386 pages. 1974.

Vol. 100: B. S. Duran and P. L. Odell, Cluster Analysis. A Survey. VI, 137 pages. 1974.

Vol. 101: W. M. Wonham, Linear Multivariable Control. A Geometric Approach. X, 344 pages. 1974.

Vol. 102: Analyse Convexe et Ses Applications. Comptes Rendus, Janvier 1974. Edited by J.-P. Aubin. IV, 244 pages. 1974.

Vol. 103: D. E. Boyce, A. Farhi, R. Weischedel, Optimal Subset Selection. Multiple Regression, Interdependence and Optimal Network Algorithms. XIII, 187 pages. 1974.

Vol. 104: S. Fujino, A Neo-Keynesian Theory of Inflation and Economic Growth. V, 96 pages. 1974.

Vol. 105: Optimal Control Theory and its Applications. Part I. Proceedings 1973. Edited by B. J. Kirby. VI, 425 pages. 1974.

Vol. 106: Optimal Control Theory and its Applications. Part II. Proceedings 1973. Edited by B. J. Kirby. VI, 403 pages. 1974.

Vol. 107: Control Theory, Numerical Methods and Computer Systems Modeling. International Symposium, Rocquencourt, June 17–21, 1974. Edited by A. Bensoussan and J. L. Lions. VIII, 757 pages. 1975.

Vol. 108: F. Bauer et al., Supercritical Wing Sections II. A Handbook. V, 296 pages. 1975.

Vol. 109: R. von Randow, Introduction to the Theory of Matroids. IX, 102 pages. 1975.

Vol. 110: C. Striebel, Optimal Control of Discrete Time Stochastic Systems. III. 208 pages. 1975.

Vol. 111: Variable Structure Systems with Application to Economics and Biology. Proceedings 1974. Edited by A. Ruberti and R. R. Mohler. VI, 321 pages. 1975.

Vol. 112: J. Wilhelm, Objectives and Multi-Objective Decision Making Under Uncertainty. IV, 111 pages. 1975.

Vol. 113: G. A. Aschinger, Stabilitätsaussagen über Klassen von Matrizen mit verschwindenden Zeilensummen. V, 102 Seiten. 1975.

Vol. 114: G. Uebe, Produktionstheorie. XVII, 301 Seiten. 1976.

Vol. 115: Anderson et al., Foundations of System Theory: Finitary and Infinitary Conditions. VII, 93 pages. 1976

Vol. 116: K. Miyazawa, Input-Output Analysis and the Structure of Income Distribution. IX, 135 pages. 1976.

Vol. 117: Optimization and Operations Research. Proceedings 1975. Edited by W. Oettli and K. Ritter. IV, 316 pages. 1976.

Vol. 118: Traffic Equilibrium Methods, Proceedings 1974. Edited by M. A. Florian. XXIII, 432 pages. 1976.

Vol. 119: Inflation in Small Countries. Proceedings 1974. Edited by H. Frisch. VI, 356 pages. 1976.

Vol. 120: G. Hasenkamp, Specification and Estimation of Multiple-Output Production Functions. VII, 151 pages. 1976.

Vol. 121: J. W. Cohen, On Regenerative Processes in Queueing Theory. IX, 93 pages. 1976.

Vol. 122: M. S. Bazaraa, and C. M. Shetty, Foundations of Optimization VI. 193 pages. 1976

Vol. 123: Multiple Criteria Decision Making. Kyoto 1975. Edited by M. Zeleny. XXVII, 345 pages. 1976.

Vol. 124: M. J. Todd. The Computation of Fixed Points and Applications. VII, 129 pages. 1976.

Vol. 125: Karl C. Mosler. Optimale Transportnetze. Zur Bestimmung ihres kostengünstigsten Standorts bei gegebener Nachfrage. VI, 142 Seiten. 1976.

Vol. 126: Energy, Regional Science and Public Policy. Energy and Environment I. Proceedings 1975. Edited by M. Chatterji and P. Van Rompuy. VIII, 316 pages. 1976.

Vol. 127: Environment, Regional Science and Interregional Modeling. Energy and Environment II. Proceedings 1975. Edited by M. Chatterji and P. Van Rompuy. IX, 211 pages. 1976.

Vol. 128: Integer Programming and Related Areas. A Classified Bibliography. Edited by C. Kastning. XII, 495 pages. 1976.

Vol. 129: H.-J. Lüthi, Komplementaritäts- und Fixpunktalgorithmen in der mathematischen Programmierung. Spieltheorie und Ökonomie. VII, 145 Seiten. 1976.

Vol. 130: Multiple Criteria Decision Making, Jouy-en-Josas, France. Proceedings 1975. Edited by H. Thiriez and S. Zionts. VI, 409 pages. 1976.

Vol. 131: Mathematical Systems Theory. Proceedings 1975. Edited by G. Marchesini and S. K. Mitter. X, 408 pages. 1976.

Vol. 132: U. H. Funke, Mathematical Models in Marketing. A Collection of Abstracts. XX, 514 pages. 1976.

Vol. 133: Warsaw Fall Seminars in Mathematical Economics 1975. Edited by M. W. Loś, J. Loś, and A. Wieczorek. V. 159 pages. 1976.

Vol. 134: Computing Methods in Applied Sciences and Engineering. Proceedings 1975. VIII, 390 pages. 1976.

Vol. 135: H. Haga, A Disequilibrium – Equilibrium Model with Money and Bonds. A Keynesian – Walrasian Synthesis. VI, 119 pages. 1976.

Vol. 136: E. Kofler und G. Menges, Entscheidungen bei unvollständiger Information. XII, 357 Seiten. 1976.

Vol. 137: R. Wets, Grundlagen Konvexer Optimierung. VI, 146 Seiten. 1976.

Vol. 138: K. Okuguchi, Expectations and Stability in Oligopoly Models. VI, 103 pages. 1976.

Vol. 139: Production Theory and Its Applications. Proceedings. Edited by H. Albach and G. Bergendahl. VIII, 193 pages. 1977.

Vol. 140: W. Eichhorn and J. Voeller, Theory of the Price Index. Fisher's Test Approach and Generalizations. VII, 95 pages. 1976.

Vol. 141: Mathematical Economics and Game Theory. Essays in Honor of Oskar Morgenstern. Edited by R. Henn and O. Moeschlin. XIV, 703 pages. 1977.

Vol. 142: J. S. Lane, On Optimal Population Paths. V, 123 pages. 1977.

Vol. 143: B. Näslund, An Analysis of Economic Size Distributions. XV, 100 pages. 1977.

Vol. 144: Convex Analysis and Its Applications. Proceedings 1976. Edited by A. Auslender. VI, 219 pages. 1977.

Vol. 145: J. Rosenmüller, Extreme Games and Their Solutions. IV, 126 pages. 1977.

Vol. 146: In Search of Economic Indicators. Edited by W. H. Strigel. XVI, 198 pages. 1977.

Vol. 147: Resource Allocation and Division of Space. Proceedings. Edited by T. Fujii and R. Sato. VIII, 184 pages. 1977.

Vol. 148: C. E. Mandl, Simulationstechnik und Simulationsmodelle in den Sozial- und Wirtschaftswissenschaften. IX, 173 Seiten. 1977.

Vol. 149: Stationäre und schrumpfende Bevölkerungen: Demographisches Null- und Negativwachstum in Österreich. Herausgegeben von G. Feichtinger. VI, 262 Seiten. 1977.

Vol. 150: Bauer et al., Supercritical Wing Sections III. VI, 179 pages. 1977.

Vol. 151: C. A. Schneeweiß, Inventory-Production Theory. VI, 116 pages. 1977.

Vol. 152: Kirsch et al., Notwendige Optimalitätsbedingungen und ihre Anwendung. VI, 157 Seiten. 1978.

Vol. 153: Kombinatorische Entscheidungsprobleme: Methoden und Anwendungen. Herausgegeben von T. M. Liebling und M. Rössler. VIII, 206 Seiten. 1978.

Vol. 154: Problems and Instruments of Business Cycle Analysis. Proceedings 1977. Edited by W. H. Strigel. VI, 442 pages. 1978.

Vol. 155: Multiple Criteria Problem Solving. Proceedings 1977. Edited by S. Zionts. VIII, 567 pages. 1978.

Vol. 156: B. Näslund and B. Sellstedt, Neo-Ricardian Theory. With Applications to Some Current Economic Problems. VI, 165 pages. 1978.

Vol. 157: Optimization and Operations Research. Proceedings 1977. Edited by R. Henn, B. Korte, and W. Oettli. VI, 270 pages. 1978.

Vol. 158: L. J. Cherene, Set Valued Dynamical Systems and Economic Flow. VIII, 83 pages. 1978.

Vol. 159: Some Aspects of the Foundations of General Equilibrium Theory: The Posthumous Papers of Peter J. Kalman. Edited by J. Green. VI, 167 pages. 1978.

Vol. 160: Integer Programming and Related Areas. A Classified Bibliography. Edited by D. Hausmann. XIV, 314 pages. 1978.

Vol. 161: M. J. Beckmann, Rank in Organizations. VIII, 164 pages. 1978.

Vol. 162: Recent Developments in Variable Structure Systems, Economics and Biology. Proceedings 1977. Edited by R. R. Mohler and A. Ruberti. VI, 326 pages. 1978.

Vol. 163: G. Fandel, Optimale Entscheidungen in Organisationen. VI, 143 Seiten. 1979.

Vol. 164: C. L. Hwang and A. S. M. Masud, Multiple Objective Decision Making – Methods and Applications. A State-of-the-Art Survey. XII, 351 pages. 1979.

Vol. 165: A. Maravall, Identification in Dynamic Shock-Error Models. VIII, 158 pages. 1979.

Vol. 166: R. Cuninghame-Green, Minimax Algebra. XI, 258 pages. 1979.

Vol. 167: M. Faber, Introduction to Modern Austrian Capital Theory. X, 196 pages. 1979.

Vol. 168: Convex Analysis and Mathematical Economics. Proceedings 1978. Edited by J. Kriens. V, 136 pages. 1979.

Vol. 169: A. Rapoport et al., Coalition Formation by Sophisticated Players. VII, 170 pages. 1979.

Vol. 170: A. E. Roth, Axiomatic Models of Bargaining. V, 121 pages. 1979.

Vol. 171: G. F. Newell, Approximate Behavior of Tandem Queues. XI, 410 pages. 1979.

Vol. 172: K. Neumann and U. Steinhardt, GERT Networks and the Time-Oriented Evaluation of Projects. 268 pages. 1979.

Vol. 173: S. Erlander, Optimal Spatial Interaction and the Gravity Model. VII, 107 pages. 1980.

Vol. 174: Extremal Methods and Systems Analysis. Edited by A. V. Fiacco and K. O. Kortanek. XI, 545 pages. 1980.

Vol. 175: S. K. Srinivasan and R. Subramanian, Probabilistic Analysis of Redundant Systems. VII, 356 pages. 1980.

Vol. 176: R. Färe, Laws of Diminishing Returns. VIII, 97 pages. 1980.

Vol. 177: Multiple Criteria Decision Making-Theory and Application. Proceedings, 1979. Edited by G. Fandel and T. Gal. XVI, 570 pages. 1980.

Vol. 178: M. N. Bhattacharyya, Comparison of Box-Jenkins and Bonn Monetary Model Prediction Performance. VII, 146 pages. 1980.

Vol. 179: Recent Results in Stochastic Programming. Proceedings, 1979. Edited by P. Kall and A. Prékopa. IX, 237 pages. 1980.

Vol. 180: J. F. Brotchie, J. W. Dickey and R. Sharpe, TOPAZ – General Planning Technique and its Applications at the Regional, Urban, and Facility Planning Levels. VII, 356 pages. 1980.

Vol. 181: H. D. Sherali and C. M. Shetty, Optimization with Disjunctive Constraints. VIII, 156 pages. 1980.

Vol. 182: J. Wolters, Stochastic Dynamic Properties of Linear Econometric Models. VIII, 154 pages. 1980.

Vol. 183: K. Schittkowski, Nonlinear Programming Codes. VIII, 242 pages. 1980.

Vol. 184: R. E. Burkard and U. Derigs, Assignment and Matching Problems: Solution Methods with FORTRAN-Programs. VIII, 148 pages. 1980.

Vol. 185: C. C. von Weizsäcker, Barriers to Entry. VI, 220 pages. 1980.

Vol. 186: Ch.-L. Hwang and K. Yoon, Multiple Attribute Decision Making – Methods and Applications. A State-of-the-Art-Survey. XI, 259 pages. 1981.